SCANNERS 6

Sixth Revised and Updated Edition

SCANNERS 6

Sixth Revised and Updated Edition

Peter Rouse and Bill Robertson

Special Interest Model Books

Special Interest Model Books Ltd
P.O.Box 327
Poole, Dorset BH15 2RG
England

www.specialinterestmodelbooks.co.uk

First published 2009
Text © 2009 Bill Robertson & Estate of Peter Rouse
Layout © 2009 Special Interest Model Books Ltd
Reprinted 2010

ISBN 978 185486 257 0

Printed and bound in Great Britain by Martins the Printers, Berwick
upon Tweed

Contents

Foreword

Here it is, the sixth edition of the 'Scanners' series of books. This edition follows in the footsteps of 'Scanners 5' and and has been updated in many sections, particularly frequency allocations together with the addition of new entries such as search and rescue frequencies, Shopwatch and Pubwatch channels, RNLI frequencies and plenty more. The Scanners and Accessories chapter has also been updated and substantially expanded, and should hopefully provide an excellent 'buying guide' for both current and second hand equipment. This book again continues the tradition pioneered by Peter Rouse in his original 'Scanners' (Edition 1) book many years ago, 'Scanners 2', and Peter commenced 'Scanners 3' but unfortunately left this world before it could be completed. I'm sure he'd be proud to know that his work is still continuing and is being enjoyed by scanner enthusiasts around the world. Much of his original information, such as radio fundamentals and antenna principles and much more has been retained. All I've done is updated this with new developments. I hope you enjoy the book, and that it adds to the pleasure of your radio listening hobby.

Bill Robertson 2009

Chapter 1
Introduction

What can I legally listen to?

Typical scanners available on the market today usually cover a very wide frequency range, and as such they're capable of tuning into a similarly wide number of radio transmissions. Like those used by the coastguard and other emergency services, civil and military airband, PMR446 'Walkie Talkies' used by both public and professionals, security guards down at your local hypermarket or shopping centre, taxis, CB operators, radio hams, pagers, local businesses, wireless radio links used for entertainment next door, taxis, private investigators, your local pizza delivery service, the list is almost endless. With the equally wide availability of scanners, many people do indeed have great fun in listening to these as a hobby in itself. Most users are sensible and keep what they hear and do to themselves, whilst others like to make a 'big thing' about what they listen to. The latter sometimes then find their knuckles get rapped! The serious side to this, of course, is that of criminals using what they hear to advantage. Fortunately more and more 'professional' users, such as security personnel, are acting on the sensible assumption that if they want their conversations to be private, they must scramble them. Others are 'open'.

I'm often asked "Surely if I don't reveal or act on what I've heard, it's OK, can't I then listen to what I want?" I make no apologies for reiterating here the caution given in earlier volumes; the aim of this book is to provide a basic understanding of the use of scanning receivers and HF, VHF and UHF communications. Contrary to popular belief this type of equipment is not solely purchased by people who have no legitimate right to listen in to certain kinds of radio traffic and who wish to illegally snoop on other people's messages. Many people from licensed amateurs to commercial and professional users buy this type of equipment for perfectly legitimate reasons. However, such people may still not fully understand how to use the scanner to its best ability. This book is aimed at all scanner users who want a better understanding of how their equipment works. In order to achieve that aim it has been necessary to include a wide range of information, some of which might be considered sensitive. Although certain bands and frequency allocations are shown, the book should not be interpreted as an invitation to listen-in - unless the appropriate licence or authority is held.

Ofcom, which is the UK's regulatory body on two-way private mobile radio communications, state, at the time this is being written (2009), that a licence is not required for a radio receiver as long as it is not capable of transmission as well. The exception to this is that it is an offence to listen to unlicensed broadcasters (pirates) without a licence, and licenses are not issued for this purpose. Although it is not illegal to sell, buy or own a scanning or other receiver in the UK, it must only be used to listen to transmissions meant for GENERAL RECEPTION. The services that you can listen to include Amateur and Citizens' Band transmissions, licensed broadcast radio stations and at sea you may listen to weather and navigation information.

It is an offence to listen to any other radio services unless you are authorised by the Secretary of State to do so.

The responsibility thus lies with you, the equipment owner to satisfy yourself that you have a legal right to listen-in to any radio transmission.

All information published here has been published before at some time, much of it by the Government. However its publication here must not be interpreted as some kind of right to listen-in to anything you wish. However, I was interested to see a document from the Department of Trade and Industry to police forces, which provided advice on seizing scanners. The document said that tuning into airband and marine transmissions could not be considered "a serious offence". Even so, I know of one individual who was formally interviewed by the police after simply admitting he used his scanner down at the local airport when plane-spotting with his son. If you are in any doubt as to what you are, and are not, allowed to listen to, you should seek guidance from your national radio regulatory body

Some cautionary advice

The law in the UK is still quite clear, in that if you want to listen to two-way radio (i.e. not broadcast transmissions) then you're limited to tuning into radio amateurs, CB operators, or whatever you're otherwise specifically authorised and duly licensed to listen to. For example, if you run a business and you've a PMR (Private Mobile Licence) then reportedly I'm told that you're reasonably OK to tune into your own business's frequency on your scanner. But not the frequencies of your competitor, for examples if you're a taxi service it's a big 'no-no' to tune into other taxi operators in your area, in fact at least one such firm owner in the UK has been 'done' for doing just this.

A Common Fallacy

It's a subject that many listeners fall down on, when they believe "you can listen to what you want, as long as you don't pass the information on". I was told this very same fallacy by a scanner listener, who was openly using his Icom IC-R2 handheld to listen to trunked radio communication on 200MHz, while I was with friends in the highlands of Scotland at a non radio-related enthusiast event. No, you can't listen to aircraft, or boats, or trains, without typically the permission of the person or organisation who's actually doing the transmitting. Neither can you get a 'licence' to listen in. This isn't helped by adverts in a UK national paper and shop windows for receivers covering the VHF airband and marine ranges stating "you can listen to aircraft" with this receiver. Admittedly, the authorities usually don't regard things like hobby-based airband monitoring a "serious offence". But try telling that to the judge. Maybe someone should, to bring about a 'test case'! Any volunteers out there?

But how about airband listening? There are plenty of low cost scanner receivers and manually tuned receivers with VHF 'Air Band' coverage included which will receive this band. As I write this you can buy a manually tuned airband receiver in the high street, complete with an airband frequency book, as a package, with a clearly marked description of "Listen to aircraft". If this isn't a clear public incitement to illegally listen then what is? As far as I'm aware, no retailer has been challenged on this by the authorities. At a number of air shows, frequencies are publicly displayed by the organisers, with airband receivers on sale at the event for visitors to buy and naturally to tune into the action. I feel you'd be hard pushed to be arrested or whatever for doing exactly that.

One person has also reported that he'd witnessed someone have their scanner confiscated by the police because they were using it inside an airport building. In another incident, my colleague who was using his handheld was challenged at the Southampton Boat Show several years ago by a Radio Communications Agency official. Once he demonstrated that it was actually tuned to the 2m amateur band and not on other frequencies (the RA thought it was being used for VHF Marine Band) all was 'sweetness and light', the RA official even asking "are these your children, would they like some badges and balloons?" - an offer which the kids promptly took up!

Confiscations

It's not a criminal or civil offence to simply possess a scanner in the UK. But if you're either caught with it blaring out messages from your local shop watch, or you admit to the fact that you're using it to monitor things you shouldn't, then it's the use you're putting it to that's illegal. One chap in my area was stupid enough to do the former, while he was having a meal in a crowded department store restaurant. Yes, he did get his 'collar felt' and rather quickly at that, and he was subsequently prosecuted. The police have powers to confiscate, for further investigation any article that they believe has been used in the committing of a crime, so that they can investigate further to see if any evidence comes to light. Like the frequencies you or someone else has programmed into the memory banks. A police officer can also seize a scanner if he has reasonable grounds to believe that it has been used for, or to aid, a criminal act. 'Going equipped' to monitor communications, for example a store's shop watch system, could be seen as intent to commit theft. So you're then likely to be rather more serious bother than just using it out of interest to listen into your local goings-on.

One suggestion I've been told of, if you're really serious about listening into aircraft and keeping ownership of your receiver, is to 'keep it discreet'. In other words don't advertise to the world that you're listening in. Using a pair of stereo earphones with the scanner in your inside pocket, and maybe making suitable physical motions to accompany personal stereo music if someone starts taking an interest, could very well work!

Back to the present world, what's changed?

So what's happened since 'Scanners 5'? The UK police and some other 'blue light' emergency services have virtually all moved over to using TETRA (TErrestrial Trunked RAdio) for their over-air communications needs. But as I mentioned in the last edition where normal FM is being used for 'Fire Ground' communication, i.e. at the incident itself usually with a vehicle-mounted FM repeater, this is now often being cross-linked to other emergency 'blue light' services using TETRA at the incident so that fire ground personnel using FM can have two-way communications with the TETRA users.

The use of VHF and UHF FM communications for 'Shop Watch' and 'Pub Watch' schemes has increased, where retail outlets and pubs/clubs in a given area have joined together in a local radio scheme to alert each other of potential trouble. PMR446 walkie talkies, offering licence-free communication typically of up to around a kilometre in built-up areas, are now extremely abundant and are used by many professionals with 'upper class' models of these radios, down to families keeping in touch with each other using budget type models which are often found on sale in the high street.

The use of the radio spectrum, particularly on the VHF and UHF allocations, is often changing, and details of these are given in relevant chapters in this book. This way, if you come across a strange-sounding signal on a given channel, a glance at the allocations section will hopefully show what that part of the spectrum is allocated to. However, in a similar manner to earlier edition I've not padded-out the book with details of what taxi firm operates on what frequency in any given area, or listings of TV and broadcast radio channels and the like.

The frequency spectrum covered by scanners has increased both upwards and downwards, and even handheld scanners now commonly cover down to below 500kHz and up to 3000MHz and above. Like most electronics, scanners are also either coming down in size, some having the area of a credit card or less if not the thinness of one, or packing a lot more features and coverage into the same size as simple scanners used to be. Some models are very sophisticated, with many now having alphanumeric memory channel tagging facilities, bandscopes, and an every-increasing number of memory channels and search ranges. But yet there are still plenty of 'easy to use' types aimed at the user who doesn't want to have to obtain a degree in IT (Information Technology) to be able to use one! 'Frequency Grabbing' scanners are increasing in number and are now available from a number of manufacturers such as Alinco and Uniden Bearcat. Here, the scanner has a 'near field' receiver system, which is essentially a sensitive off-air frequency counter which looks for strong signals, i.e. those in the vicinity of the scanner, and measures their frequency, then automatically tunes the scanner to that frequency so you hear the signal itself. There's usually also an option to automatically and store it in the scanner's memory bank for subsequent monitoring

Many people using their scanner from home are now linking it up via the remote port to their home PC. Here, the PC can be used in conjunction with a scanner for control and frequency management with appropriate software, not only to remotely control a scanner, but also for automatic signal logging, and recording of received signals to hard disk. Some radios are simply a 'black box', controlled exclusively by a PC with a 'virtual front panel' being displayed on the PC screen. Some, like the Icom IC-R1500PC can be either controlled by PC or with a plug-in hardware front panel, giving the best of both worlds.

Data communication over radio is ever increasing, with systems for automatic location (with a GPS receiver linked to the user's mobile two-way radio) now very commonplace. Not just for taxis, but delivery firms, security firms, and plenty more. Trunked communication, where 'intelligent' radios use whatever channels are available at any time from a central pool, is also increasing, and there is readily-available PC software for this, as well as an increasing number of similar 'intelligent' scanners which can automatically track some trunked systems.

Although cellphone usage with SMS text facilities is very common, these can't be used everywhere and paging systems are still very much in use, for emergency call out, as well as systems such as traffic congestion data. Both numerical and alphanumeric paging systems are in use, and plenty of freeware and shareware PC software programs are available to decode these, with the PC linked to your receiver through the PC's sound card. Software for decoding other signalling, such as CTCSS (sub-audible tones) and DTMF (Dual Tone Medium Frequency, i.e. 'Touch Tone') is readily available as are off-air weather image transmissions from low-earth orbiting and geostationary satellites.

The use of encryption, i.e. speech scrambling, is increasing as radio users are increasingly aware that their transmissions may be overheard by others. But scanner manufacturers haven't been slow to respond, and there are a number of handheld and base scanners with the ability to decode the simpler inverted speech methods of scrambling, there are also readily-available PC programs to handle this.

The two-way radio communications scene will, as ever, continue to change, and I hope this book with help you to get the best out of your receiver, whether you use it for hobby or professional monitoring purposes.

Chapter 2
Understanding radio

If you are new to the hobby then have a browse through this chapter, because it answers the queries most frequently raised by scanner owners who are not familiar with HF, VHF and UHF communications. First of all let us look at what we mean by this. The entire radio spectrum stretches from Long Wave upwards and as we go higher in frequency so different parts of the spectrum are given different names. The best way to imagine it is to think of a very long tuning scale on a radio. At the left side we have long wave, then comes medium. From then on we talk in terms of frequency rather than waves. Note that there is a fixed relationship between wavelength and frequency and it is always possible to determine the wavelength of any given frequency or vice versa. So a look at **Table 2.1** will show how things progress beyond medium wave. Next comes Medium Frequency (MF), then High Frequency (HF) which is also known as Short Wave. From 30 MHz (Megahertz) to 300 MHz we have Very High Frequency (VHF) and between 300 MHz and 3000 MHz we have Ultra High Frequency (UHF). Look again at Table 2.1 and you will see reference to both kHz (kilohertz) and MHz and it is useful to understand the relationship between the two. They are simply measurements and the kilo and Mega parts are the same as those applied to measuring

Table 2.1 Radio frequency spectrum, each division as a frequency range

Frequency division	Frequency range
Very low frequency (VLF)	3-30 kHz
Low frequency (LF)	30-300 kHz
Medium frequency (MF)	300-3000 kHz
High frequency (HF)	3-30 MHz
Very high frequency (VHF)	30-300 MHz
Ultra high frequency (UHF)	300-3000 MHz
Super high frequency (SHF)	3-30 GHz
Extremely high frequency (EHF)	30-300 GHz
No designation	300-3000 GHz

k = kilo = x1,000. M = Mega = x1,000,000. G = Giga = x1,000,000,000

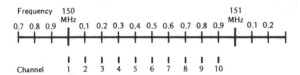

Figure 2.1 Channelising. Any section of the radio spectrum can be divided up and each spot frequency given a channel number

metric length and weight. Kilo means a thousand times and Mega means a million times. So 1000 kHz is exactly the same as 1 MHz. Understanding this will help you grasp the idea of what is known as tuning or stepping rates on your scanner. Scanners are not tuned like an ordinary radio. If you want to alter a frequency up or down you will have to do it in small jumps. Sometimes the scanner will only tune in 5 or 10 kHz increments but those designed specifically for the British market will often have 6.25kHz and 12.5 kHz increments because this is a common British channel spacing and channel offset.

Channels

Channelizing is something, which often confuses newcomers, and yet it is quite simple. We can take any part of the radio spectrum and divide it up into small blocks. A look at **Figure 2.1** will show how this is done. In most countries, the authorities determine how the channel spacing will work. It is common in Britain on the bands used for two-way radio and Private Mobile Radio (PMR) to use channelizing, rather than allocate frequencies to users at random. Typical spacing in Britain is 12.5 kHz, sometimes with a 6.25kHz offset but in the USA and other countries the spacing for two-way non-aircraft radio is usually 10 kHz. This can present problems for some scanner owners who have receivers designed for the American market, because they cannot tune exactly onto the right frequency of many British channels. For example it is not possible to enter the frequency 144.7625 on some scanners although dialling-in 144.760 may be close enough to hear the signal: it varies from one model to another. It is a point to watch when buying a scanner. Also, early scanners covering civil airband on VHF (118-137MHz) will only tune this in 25kHz increments, but although this spacing is still widely used some channels have progressed to 8.33kHz spacing, i.e. a third of 25kHz, so it's important to look for this if you're interested in scanning through the airband channels.

Distance

How far will a signal travel? Why can I hear an aircraft over 30 miles away but not my own airport, which is just over the hill behind my house?

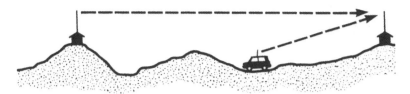

Figure 2.2 *Ground wave. Typically line of sight at VHF and UHF.*

These are typical of the questions from newcomers. VHF and UHF signals travel nowhere near as far as medium and long wave signals. In fact they only travel over what is known as "line of sight"; typically 20 to 30 miles. In practice the distance can be far less or much greater depending on circumstances. One factor is the power output of the station we are listening to. A base station with between 25 and 50 Watts output power may come in loud and clear over a given distance but a small walkie-talkie with only a fraction of a Watt output might not cover the same distance. You should also bear in mind that VHF and UHF signals are easily blocked by obstructions such as high buildings, hills and even trees. There are also certain weather-based conditions that can cause signals to travel massive distances. One of these is known as Sporadic-E. VHF signals do not normally bounce back to earth off the ionosphere (the effect that causes HF signals to travel great distances) but on occasion the E-layer of the Ionosphere thickens up and the signals do bounce back. This is when VHF FM broadcast stations can sometimes come through from around 1000km away, although the effect usually lasts no more than a few tens of minutes. Another effect is known as Tropospheric Ducting. This happens when cold and warm air streams meet at about 2

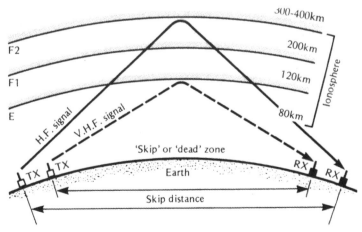

Figure 2.3 *Sky wave. Note that VHF signals will not normally skip as far as HF signals.*

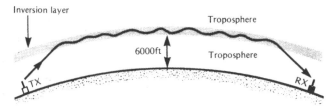

Figure 2.4 *Tropospherical ducting 'Tropo'.*

kilometres above the earth's surface. A conductive layer is formed that can act as a sort of pipe for signals and send them considerable distances (several thousand kilometres in some cases). This effect is often associated with high barometric pressure and fog and the effect is often seen in the summer, when VHF FM broadcast stations can come in from long distances away, although not as strongly as Sporadic E but the effect usually lasts for up to an hour or more.

Modes

There are several ways of imposing speech (and music, video, and data) onto a radio signal. For speech the three common methods are frequency modulation (FM), amplitude modulation (AM) and single sideband (SSB). You don't need to understand the technical difference between the modes but you must appreciate that they are different and your scanner must have different circuits to cope with the different modes. For instance all transmissions in the international VHF aircraft band use AM, and if your scanner is tuned to a station but FM mode is selected, you will hear either nothing or at best a very distorted signal. Britain is one of the few places in the world now where some two-way radio users still use AM mode for communications outside the aircraft band. For instance in a given area you are likely to find both AM and FM transmissions being used in a PMR band. Most countries and particularly the USA exclusively use FM with its many advantages. If you want to pick up a wide selection of transmissions then you need a scanner that allows you to select FM or AM regardless of the frequency you are tuned to. Many scanners now allow you to store a frequency into memory along with the mode. However, you will find several scanners on the UK market, which do not allow you to do this. Notably, scanners designed for the American market have no FM/AM select button and instead automatically switch to FM on any frequency except the air band when they switch to AM. Single sideband is a rather specialised mode and is divided into upper and lower sidebands. To resolve SSB your scanner needs special circuitry and this is only found on a handful of models although the number is increasing. The mode is used by amateurs on the VHF and UHF bands, as well as a wide range of utility stations on HF. Models of scanners with HF coverage as well as SSB reception facilities can thus bring in an extra area of interesting listening.

Simplex and duplex

A question that's often raised by newcomers to VHF/UHF communications is why they can only hear one of two stations that are communicating with each other. This is because they may be using either split frequency simplex or duplex. First though, let's start at the beginning and look at the simplest arrangement that is called simplex. A typical example of single frequency simplex is the international aircraft band between 118 and 137 MHz. A control tower for example may be operating on 119.950 MHz and the aircraft will also transmit on exactly the same frequency. Assuming that both are within reception range of your scanner then you will hear both sides of the conversation on the single frequency of 119.95 MHz. However, some transmissions use a pair of frequencies with the base transmitting on one frequency but with mobile stations on another, this is split frequency simplex. In this instance the transmitters and receivers will have to be tuned to the appropriate frequencies. The scanner user will need to enter both frequencies into the scanner and switch between them to hear both sides of the conversation. Some upper-market scanners have this 'duplex frequency listen' ability by the press of a button.

Repeaters

Some users who need to obtain greater range for their communications rely on what are known as repeaters. These transmitters/receivers are normally situated on high ground and have the added advantage of allowing

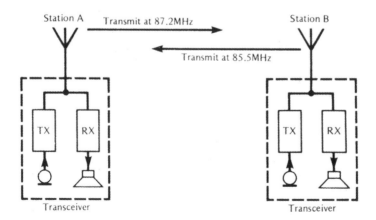

Figure 2.5 *Split frequency simplex.*

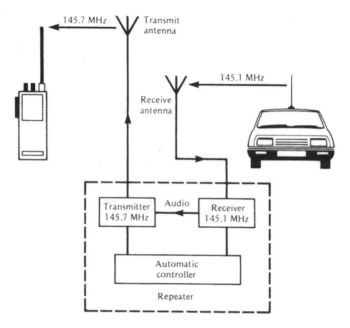

Figure 2.6 Repeater. Automatic turn-on and turn-off. In the case of 'talkthrough' the switching controller is operated manually.

one mobile unit to talk to another. Where the police use such systems this is known as 'Talkthrough'. Repeaters work on split frequencies. In other words the 'input' or receiver frequency is usually several Megahertz away from the transmit or 'output frequency'. The transmitter normally is switched off but when the repeater's receiver picks-up a signal it then activates the transmitter and the audio output of the receiver is fed to the input of the transmitter and so rebroadcasts it. In practice the receiver will not respond to just any signal but also requires a valid tone. On some amateur repeaters this can be a simple audible tone burst of 1750Hz that is detected by a circuit in the repeater. On commercial systems as well as virtually all UK amateur repeaters sub-audible tones are used (known as CTCSS) and increasingly continuous digital codes (DCS) on commercial systems. Some commercial repeaters may also be programmed to respond to several different tones, this is the case where more than one operator is using the repeater. The system is often referred to as a 'Community Repeater' and can be used by several companies or organisations. Each has its own CTCSS tone / DCS code and the access circuits are not only fitted to the repeater but also the individual mobile or handheld transceivers. In that way the various users only hear calls from their own base station or mobiles. In order to monitor calls on a repeater system you only need to tune to the repeater output frequency. A variation on this theme is what are known as trunked systems. This is where a network of repeaters is used and controlled by computer. A typical network may cover most of the country and so a base station can call its mobiles as long as they are within range of any of the trunked repeaters.

You can read a more in-depth description of the above, plus details on trunking, selective calling, etc. in the 'Radio Systems Explained' chapter later on in this book.

Chapter 3
The Hardware

So what is a scanner? The simple answer is that it is a radio receiver that's capable of tuning into radio frequency channels that you won't normally find on your ordinary domestic radio. What also sets it apart from a manually tuned radio is that it will also look automatically through dozens or even hundreds or thousands of channels looking for radio signals. Unlike broadcast transmissions those frequencies will only carry occasional conversations and so the scanner constantly sweeps through its memory channels and only stops when it finds an active one. Once the transmission ends the scanner then resumes sweeping its pre-programmed channels looking for another active one. The required frequency will usually be entered on a keypad, with the ability to store the frequency in a computer style memory and with the capability of many channels being stored. The simplest scanners may only have the capacity to store up to around 50 or 100 channels, but more sophisticated scanners also allow such things as mode and a short alphanumeric name to be stored along with the frequency, so that once the scanner is put to work it automatically switches to the correct mode for the frequency concerned and reminds you of what you've programmed the channel to receive. All of this is made possible by a simple and compact microprocessor within the scanner and in most instances it allows for even greater sophistication. For example it is usually always possible to set the scanner searching between a lower and upper frequency limit so allowing you to discover new channels that may be in operation in your area. Some scanners also have an 'auto store' facility, where you can set it searching over a given frequency range and it will automatically store active channels for you into its internal memory. Finally, a breed of scanners are equipped with a 'Close Call' or 'Transweeper' facility, the name depending upon what the individual manufacturer calls it. Here the scanner can automatically check for the presence of local signals, automatically tune the scanner to the frequency it's found, providing of course it's within the frequency coverage range of the scanner, to let you listen to what it's found. Besides this, you can also get it to again automatically store the frequencies it finds into the set's memory.

You get what you pay for

Newcomers to the scanning hobby often ask what the difference is between a cheap scanner and another costing a small fortune. The answer is facilities. Some of the Realistic brand handheld scanners are probably the cheapest and simplest receivers on the market. The lowest cost type has just 50 memory channels, limited frequency coverage, and is FM only. I must stress that, despite that, it may well fill the needs of someone who for example only wants to monitor a handful of marine channels. At the other end of the scale are the Icom IC-R9000 and AOR AR-5000 with very wideband coverage and good front-end filtering. This latter facility is often not understood by scanner owners and its worthwhile looking briefly at what it involves.

In any given area there will be a vast number of radio signals travelling through the airwaves - everything from powerful broadcast transmissions to weak communications signals. The less expensive scanners will simply amplify all these signals in what are known as the Radio Frequency, or RF, stages of its circuitry before passing the signals to those parts of the circuit that at any given moment select the required signal which eventually passes through to the loudspeaker. In practice, the RF amplification is usually divided into at least two separate stages, one for the lower frequency coverage and one for the upper. However, what can happen is that strong signals that you do not wish to hear can interfere with the scanner circuitry and block out those signals you are trying to listen to. In severe cases, several undesirable effects may be introduced into the scanner's circuit. Typical examples are blocking, reciprocal mixing and intermodulation distortion. One of the most obvious effects is where you are listening to a transmission and it suddenly seems to stop for no apparent reason. What has probably happened is that a very strong signal has appeared on a slightly different frequency and has desensitised the scanner causing the receiver to mute. Another undesirable effect that can occur is when two or more strong signals appear to upset the scanner's mixing circuits. This can often be identified by a signal suddenly breaking through on a frequency which you know is incorrect for that transmission. For example, at some locations in built-up areas with plenty of PMR and pager transmitters, a certain mixture of transmissions together with, say a local radio FM broadcast can cause the latter station to appear on some channels of your scanner. Of course the signal is not really there, the internal mixing circuits in the scanner are being overloaded and cause this effect. The way that the above problems can be overcome is to have more selective circuits in the scanner's RF stage. Tuned filters ensure that only the desired signals are passed to the tuning and mixing stages. As the scanner steps through each frequency in its memory, so the correct filter is selected or the RF stage is electronically tuned to peak at that frequency. Naturally all of this makes the scanner more complex and so more expensive. The added components also make the set bulkier and that is why most tiny hand held scanners do not incorporate much in the way of front-end filtering. Owners of handhelds now know why their equipment often suffers problems when connected to an outside aerial such as a discone or a wideband amplified vertical type. The RF stages are presented with stronger signals than they are designed to cope with.

Switches, knobs, sockets and things that go click and squeak

The average scanner presents a bewildering array of facilities for anyone not used to communications equipment. However, most of the facilities are fairly straightforward although some handbooks do leave a lot to be desired in the way they explain them. Let's take a look at some of the things you will find on most scanners.

Squelch or mute

This is an electronic switch that cuts noise to the loudspeaker when no signal is being received. It also serves the important function on a scanner of telling the set that transmission on that channel has ceased, so please start scanning again. You need to set the squelch by turning it to the point where, with no signal being received, the loudspeaker cuts out and there's no longer a 'hiss' from the loudspeaker. It is important to note that you should never advance the squelch any further than is necessary. The more you turn it, the more you desensitise the scanner to the point where it will only respond to very strong signals; the weak ones will be missed. Some scanners offer a variety of types of squelch including one where even if lack of background noise opens the squelch, the scanner will continue to hunt through its memories unless there is sound such as a voice present. In fact this is often known as voice-scan or audio-scan. My personal experience is that this facility rarely works very well. One further facility is Scan `Delay'. Sometimes you do not want the scanner to resume scanning until you have heard any replies to a transmission. Scan Delay allows for a small lag before scanning commences so that you can catch the reply. Only when the scan delay period is exceeded with no further transmissions does the circuitry recommence scanning. Some top-of-the-range scanners also have a CTCSS or DCS decode facility, where the squelch will only raise when the correct tone or code is received. This can be especially useful when you're listening out for a specific user on a channel that's shared by many others, as channels often are.

Search

With this facility, nobody's frequency is safe from you. Let us assume that you have a sneaking suspicion that interesting transmissions are taking place in your area between 160 and 170 MHz. Now because of the intermittent nature of communications transmissions it can be like looking for a needle in a haystack to discover which actual frequencies are in use. Get the scanner to do the hard work. On virtually all scanners it is possible to select a lower and upper search limit and leave the scanner to sweep between those two frequencies. Naturally, it will stop when it finds something. A further refinement on this appears on higher-class scanners when the scanner can automatically also store the frequencies it finds into a spare memory bank for you. This means you can even go off for the day and come back and see which frequencies were active.

Stepping rates

Searching is always done in conjunction with stepping rates. Unlike a conventional radio, a scanner does not have infinite degrees of tuning. In other words as it goes up or down in frequency, it does it in 'small hops'. The steps available on the scanner are important, because they may prevent you from properly tuning into the correct frequency. In the UK we use 12.5 kHz channel spacing, and on some frequency ranges such as 446MHz on the aptly named PMR446 licence-free range, a 6.25kHz offset from these channels, and that can be a problem for scanners that can only tune in 5 or 10 kHz steps. What will usually happen is that you try to enter the correct digits and press the 'enter' key, the scanner's circuit will round the number up or down to suit its available stepping rate. In the worst case this means that you are slightly off-tuned from the frequency you want, and you'll either not hear the required signal or it will be distorted. At best, the scanner's filters may be fairly broad band and you will still receive the signals with little difficulty. Again, the above should be noted when choosing a scanner, and you should be aware that the previously mentioned units designed for the American market may well not have 12.5 kHz stepping rates. VHF airband is now using 8.33kHz spacing, so if you're interested in civil airband monitoring then make sure your scanner has these frequency steps available for tuning and searching.

Priority channel

Some scanners include what is known as a priority channel and indeed may have more than one. This channel or channels are scanned more frequently than the others are and the idea is that the most important frequencies are assigned to the priority positions. Typically the scanner will very briefly jump to the 'priority' channel every couple of seconds, check whether there's any activity and if so will stay on that channel, otherwise it'll jump back to the channel you were otherwise monitoring.

Bandwidth and mode

Many scanners can receive a number of modes, the most common being FM or NFM, WFM, and AM. NFM is usually used to indicate 'Narrowband FM' with WFM as 'Wideband FM', to differentiate between the two. The difference is in the receiver bandwidth filtering, NFM is used for two-way communications whereas WFM is used for high-quality broadcast sound, i.e. on the 88-108MHz 'Band II' broadcast band and for UHF analogue TV sound. Weather satellites are a 'half-way house' though, and use a wider FM deviation greater than that used for two-way communication, but narrower than that for WFM. Receivers with NFM (or just FM is it's often called in receivers dedicated to communications use) need to reject signals 25kHz, and preferably 12.5kHz, away from the tuned frequency. Scanners such as those from AOR, Icom and Yupiteru do normally have such 'tight' filtering, and offer good results on the communications bands, but these are often too narrow for good results on weather satellites. In my experience, those from Realistic/Tandy/Netset/Uniden Bearcat etc. usually have wider filtering to primarily accommodate FM use in the US rather than the narrower FM deviation used in Europe. Although this might not give as good rejection of 12.5kHz spaced signals, the results on weather satellite reception in NFM mode are somewhat better.

On some more up-market scanners a further facility is single sideband (SSB) and depending on the scanner very narrow-band circuits may be employed for this mode. SSB uses a system where the carrier of a double-sideband AM signal together with one of the sidebands is removed before being transmitted, and the 'missing' carrier is re-inserted in the receiver. As each of the sidebands, one being the Lower Sideband (LSB) and the other being the Upper Sideband (USB) are in fact a 'mirror image' of each other, you'll get fully readable reception from either.

Tuning in an SSB signal requires a degree of practice though. It'll initially sound like a 'Donald Duck' type audio, and very careful tuning, in fine 100Hz or 50Hz tuning steps, is needed. After approximately tuning to the signal, switch to the fine tuning steps if you haven't already done so and carefully tune up and down until the tone of the received speech sounds as natural as possible. If you can't get any intelligence at all even though it seems to be speech you're receiving, then switch from LSB to USB or USB to LSB and try again. Commercial users inevitably use USB for speech, whilst amateurs use LSB on radio frequencies below 10MHz and USB on those above 10MHz.

RF Attenuator

This is a switch (it may be marked Local/DX) that allows you to desensitise the scanner. Remember the earlier comments about strong signals upsetting the tuning stages, this facility will reduce or can even eliminate the problem. Unfortunately the side effect is that it will also cut down the weaker signals so that you may no longer be able to be able to hear them. Even so, in extreme cases it can be useful and at least allow you to hear some transmissions, whereas the scanner may otherwise permanently lock-up because of overload. If your scanner has HF coverage and you connect an external HF antenna, you'll invariably need to use the attenuator on some frequency ranges.

Tape control

This is coupled to the squelch circuit and will automatically switch a tape recorder on when transmissions are received. It is usually only available on the more expensive base models. However there are several automatic recording facilities available for PCs, which will automatically start and stop recording when audio is present on the signal. Thus the absence of tape control isn't really too much to worry about if you'll be using your scanner at home.

Memory backup

One point to note with some early scanners, although usually not recent models, is that the memory that stores all the frequencies in these sometimes utilises what is sometimes known as volatile RAM (random access memory), or VRAM. In order for the information to be retained whilst the scanner is switched off it is necessary to keep a small amount of power running to the memory chips. This is usually done by having an internal lithium 'button cell' battery to provide memory power. These cells will not last forever and from time to time will need to be replaced. It is always a good idea to keep a list of all the frequencies and modes you have stored in each memory if you have such an earlier scanner, providing of course you are allowed to listen to these, because the day may well come when you switch on and find that everything has gone blank. It is worth noting that sometimes the loss of backup power can also cause the scanner to behave in an odd way. If it has lost all memory and fails to operate correctly even when all data has been re-entered then it may still mean that the system needs to be reset by the inclusion of a new backup cell. Newer scanners sometimes employ a better solution, that of EAROM (Electrically Alterable Read Only Memory), which stores all your changes electrically into a memory IC which doesn't need a backup for information retention. But note here that sometimes the EAROM is electrically written only the instant you manually switch the scanner off each time. So, if you let your scanner's battery pack go flat, or disconnect it, whilst the scanner is still switched on then you may lose the information you've changed, i.e. new frequencies stored or altered, during that operating session.

Computer control

Many scanners, particularly base types but also an increasing number of handheld models, currently have a dedicated socket for computer control, although in some cases you'll need an optional interface to connect the scanner to the RS-232 or USB port of your PC. As well as commercial software, there is an increasing amount of shareware software available on a 'try before you buy' basis. Another use for the computer control port is to couple a 'frequency searcher' such as the Optoelectronics Scout to the scanner, where when the searcher finds a nearby frequency it can automatically tune your receiver to that frequency.

No Scanner but still want to listen?

You don't necessarily need to own a scanner receiver if you want to have a listen around on the airwaves. The Internet is evolving in an explosive way, and one use of this has been for audio-based relay. There are multitudes of web sites in existence which carry live audio from various police and airport radio systems in the US and many other countries. A mouse click on the appropriate hyperlink then results in live received audio from the service in question coming from your multimedia PC's speakers. An extension of this is scanner enthusiasts' sites, with a remotely controlled scanner receiver and aerial system ready installed and controlled by the host computer linked to the Internet. Here you can enter the frequency or frequencies you wish to listen to directly, or select one of a number of frequencies from a 'menu' list at the remote end.

Choosing your first, or next, receiver

It may seem an obvious question, but there's little use in looking around for a receiver with the widest coverage possible and plenty of modes if you're only interested in say, the VHF and UHF airband ranges together with ground support frequencies ranges, and you've no interest whatsoever in HF for example. But then, the 1000-3000MHz range is becoming more and more interesting, especially if you're into advanced interests such as video transmissions, both from the ground (i.e. video intercoms and cordless video links) and from overhead helicopters sending live video back to the ground. Decide what your current interests are, and if you're just 'starting out' I'd advise against spending a fortune on the latest hyper-expensive receiver with it's equally hyper-complicated controls and buttons.

Frequency Ranges and scanning steps

Make sure you buy a receiver that suits your listening interests. For example, many budget scanners offer airband coverage on VHF AM in their descriptions, but they don't cover the UHF military airband range. Others do cover that range, but some can't be switched from FM to AM on bands aside from 108-137MHz. If you're interested in terrestrial narrowband FM reception in the UK, most services operate on 12.5kHz steps, sometimes with 6.25kHz offsets, so ensure your scanner can accommodate this. Some scanners only allow 5kHz, 10kHz, 15kHz, or 25kHz steps, and although 5kHz steps will usually bring you 'near enough' you'll find it takes a two and a half times longer to search across a given range in the inappropriate 5kHz steps rather than in 12.5kHz increments. Some services, such as 'eye in the sky' traffic reporter links also operate with a 6.25kHz offset so this step could also be useful. You might also consider checking for 8.33kHz steps for VHF airband listening in Europe.

How much do you want to pay?

In general, although there are some exceptions, the more you pay for a receiver the more features you're likely to get as well as hopefully a better technical performance. If you buy a budget handheld and connect this to a rooftop antenna, particularly an amplified type, don't be surprised to find it 'falls over' with breakthrough from unwanted strong signals on other frequencies when you're trying to listen to a wanted signal. A higher price usually gets you a higher number of available memory channels, several search banks, auto-storage of active frequencies, a switchable attenuator (very useful in busy radio locations to reduce breakthrough), and maybe multi-mode reception to include SSB (Single Sideband, LSB and USB) and CW (Morse code). Few VHF/UHF commercial services use these modes, although if you're interested in listening to amateur radio DX (long distance) stations and contests, it can open up a specialised new world of listening.

HF Coverage

Many receivers offer extended coverage to include the HF (High

Frequency) range, i.e. 3-30MHz and maybe even lower frequencies, and SSB reception is essential here for utility station monitoring although AM is fine if you only wish to listen to broadcast stations. But don't expect to hear much on a set-top antenna or a discone on HF, you'll need a length of wire instead, preferably mounted outdoors and well in the clear. The usual adage here is "the longer the length of wire the better", but beware as many of the lower priced wideband receivers, and I'm talking of below £300, won't give the performance on HF that you'll get with a similarly priced dedicated HF-only receiver. In other words, connect a good outdoor purpose-designed HF antenna, and you'll often overload the receiver with unwanted off-frequency strong signals to the detriment of what you're actually trying to listen to. If your primary interest is HF, i.e. short wave utility, amateur, and broadcast stations, then spend most of your available budget on a dedicated HF receiver and somewhat less on a further receiver to cover VHF and UHF.

Portable or Base

A handheld portable scanner is very useful in that it can double for use at home or outdoors, and will often cost less than a base receiver having similar frequency coverage. A handheld will invariably have a set-top BNC or SMA coaxial antenna connector so that you can, if you wish, use the receiver at home or out mobile with an external antenna connected in place of the set-top helical. Base scanners are, however, usually easier to operate as there's more room available on the front panel for additional buttons and controls, together with a larger frequency and channel display, and often, but not always in the case of 'budget models', give better technical performance than a handheld in terms of their ability to receive weaker signals and more importantly reject unwanted signals. You pay your money and take your choice.

Mobile Scanning

Many enthusiasts have their listening post at home, sometimes supplemented by a handheld scanner for use outdoors. But if you're also using your handheld scanner in your car, you could possibly be getting poor results. If you've a permanent installation of, say, a mobile scanner with good antennas etc. then I won't be teaching you anything new. But if you'd like to 'have a go' and improve your listening possibilities, then read on! Firstly, consider the mounting. If you simply place your handheld scanner on the passenger seat, then invest a few pounds on an air-vent belt-clip mount or even a 'universal adjustable' cell phone holder if your scanner is an appropriate size. This will raise your scanner's speaker and controls to an easily usable location, where you'll reduce the eye-movement needed from the road ahead to see what's happening. It'll also give you somewhat better receive audio. Finally, the antenna. If you're using a dash mount then the set-top antenna may well have a 'clear view' through your car windows. But you'll invariably be getting noise from dashboard electronics as well as some attenuation, especially from heated and/or tinted screen coatings. A small external antenna here will reap great benefits, and a discrete magnetically mounted or window-glass mounted antenna will considerably increase your listening range. If you're interested in PMR (Private Mobile Radio) then a 2m/70cm dual-band amateur radio antenna could be ideal. If instead you were mainly interested in civil and/or military air band, then a quarter wave on 120 MHz (with three-quarter wave resonance at 460MHz for the military air band) would be better. This is a simple whip element of around 59-60cm in length. A hint here - buy a 'mini mag-mount' and cut the whip element to this length for air band, or to around 40cm for VHF and UHF PMR use, as around 40cm will be resonant on both 150MHz and 450MHz.

Mobile Scanner Mounts

An obvious use for a handheld scanner is to provide a different sort of in-car entertainment to a CD/MP3 player or broadcast radio. By using a cellphone holder, a handheld scanner can easily be mounted on the dashboard with the set-top antenna typically at window level, and thus being able to 'see' incoming radio signals. MFJ indeed make a dedicated clip mount holder that should match most handheld scanners, although I've usually used a Watson QS-200 Air Vent Fitting Mobile Mount for my scanner. This clips onto the air vent grille with the scanner's belt clip in turn clipping into the holder. An alternative is a Watson QS-400 Adjustable Mount, this again clipping into the air vent grille, hence no holes needed in your dashboard. However, as well as possibly adding an external speaker, one of the best improvements you could make to this is by adding an external antenna. Some car windows have metallic-based tints built in, most rear windows in any case are heated with metal elements effectively screening radio signals from behind, and a growing number of front windows are also of the heated variety. Don't think that adding an external antenna means drilling a hole on the roof or permanently attaching something using, say, a gutter clamp as used for many amateur radio whip antenna mounts. That is indeed if your car has a gutter, many nowadays don't! In the past I've used a dedicated glass-mount antenna and found it to be nicely discreet, especially as its matt black in colour and quite inconspicuous. I can however see that some scanner users would like to have a permanently fitted antenna, and here a small magnetically mounted type could fit the bill. Forget those old large magmounts, nowadays a tiny 'rare earth' magnet can be incredibly powerful for its size, and one of these with a thin matt black whip antenna also looks very discreet, with the advantage that you can remove it in just a couple of seconds. Several amateur radio and scanner dealers have both 'straight' whip and multi-band whips available.

Audio when mobile

Going back to the audio side of things, an alternative to an extension speaker is one of the commonly available audio interfaces for personal stereos, usually a wireless in-car FM transmitter. You plug this into your scanner's earphone jack and the audio is fed to a low power FM transmitter, with just a few metres range, which you can tune to on your car radio. Here you'll listen to scanner audio through your car's in-car entertainment speakers, which will be a lot more audible than a small handheld scanner speaker will! Another option, especially if you don't want the adjacent car occupants next to you in the car park listening to what you're listening to, is a 'dummy' compact cassette which slots into the cassette slot of a car stereo, and is again fitted with a lead which in turn plugs into the earphone socket on your scanner. Your car radio and its speakers then again act as the audio output. Whichever you use you may find that you only get audio on one side of the stereo pair if your scanner has a mono earphone jack rather than a stereo one. Here, a quick replacement with a mono 3.5mm jack plug, if you or a friend is handy with a soldering iron, will fix this. Alternatively you could cut the lead somewhere near to the existing stereo plug and re-wire the lead so that both 'inners' of the stereo pair go to the tip of the stereo plug, rather than one going to the tip and the other to the 'ring' (the middle connection of the three on the plug). But then just with audio on one side of the car's speakers is often better than trying to listen on the tiny internal speaker on the scanner.

Chapter 4
Antennas

Nearly all scanners are supplied with some sort of antenna to get you going. With the exception of hand held types, that's about all the telescopic ones are good for which are often supplied with base scanners. To get the best out of your base or mobile scanner you will need to invest in a proper antenna ('Antenna' is used here for worldwide readers and is essentially the same as the correct British word of 'Aerial'). You will also need to connect between antenna and scanner using the proper type of cable. It can never be stressed enough that an outside antenna mounted as high and as well 'in the clear' as possible is an essential part of the receiving system. We are dealing with low-level communications signals and so we need to gather every bit of them that we can. A proper antenna system will pull in signals that you will never hear just using a telescopic whip attached to the scanner. If you use an external antenna then note that you must disconnect the telescopic, as leaving it in circuit will upset the input impedance to the equipment.

VHF/UHF antennas

The antenna needed for a scanner must not be confused with the long wire types used for other types of radio reception. This is a very specialised area and the antenna and cable must be well sited and in good condition. Signals are subject to the high losses at these frequencies and even minor defects in the system will almost certainly degrade performance. VHF/UHF antennas fall into two broad categories; broadband and narrowband. The former performs well across a wide range of frequencies and the latter is suited to one particular band. The narrow band type should not be ignored as it has its uses, for instance where the air and marine bands are concerned it may well perform much better than a broadband type. The broadband antenna is in fact something of a compromise. It is also advisable that you understand about polarisation. This simply means that transmission or reception takes place via elements that are either vertical

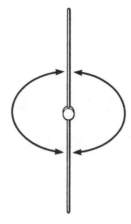

Figure 4.1 *Vertical polarisation.*
Pick up pattern is from all
directions.

to the ground or horizontal. Most communications take place using vertically polarised antennas and for optimum reception the scanner antenna should also be polarised the same way. The occasional exception to the rule is with some amateur transmissions and commercial communications between two fixed points. The disadvantage of using horizontally polarised transmission for most communications is that horizontal antennas are usually directional to one degree or another. This is hardly desirable where mobile communications are concerned because the signals can come from any direction.

Narrow band antennas

The simplest narrow band antenna is the quarter wave element or whip, and it is possible to trim a whip to match the frequency in use. We must note that the input impedance to the scanner will be the standard 50 Ohms. Using a standard formula of "75 divided by the frequency in MHz" to give the length of a quarter wave in metres, we can work out what the

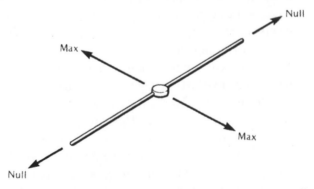

Figure 4.2 *Horizontal polarisation. The aerial is sensitive in two directions*
but picks up very little end-on.

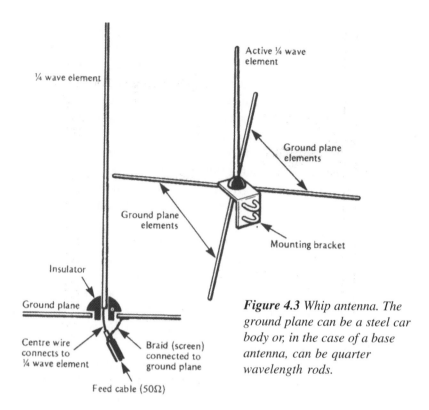

Figure 4.3 Whip antenna. The ground plane can be a steel car body or, in the case of a base antenna, can be quarter wavelength rods.

length of a whip should be, but note that we must reduce it by 5 percent to produce the 50 Ohm match. If we mismatch then some signal will be lost. Note that the antenna will work at its best and is 50 Ohm matched only when it is mounted above a ground plane such as a vehicle body or rod radials. Although this is technically a narrow band antenna it will provide good receive performance over several MHz and is ideal for marine or air bands. This type of antenna can easily be made from the parts of old TV or broadcast FM radio antennas, or for indoor use even from parts of wire dry-cleaning type coat hangers. A further simple antenna is the dipole (**Figure 4.4**) with each 'leg' being a quarter wave whip. A refinement is the 5/8th whip, which can provide a few decibels of gain and so give a slightly better performance. You may have seen this type of antenna mounted on two-way radio equipped vehicles where a long whip terminates in a small black coil close to the mounting unit.

Broadband antennas

These are by far the most popular antennas for scanner use and they can be subdivided into active and passive types. The most popular is an antenna known as a discone, and despite what some advertisements claim by suppliers this is a vertically polarised antenna even though the top elements are mounted horizontally. It is what happens between the top and bottom elements that determines the polarisation. The discone is a reasonable compromise for broadband use although it must be noted that at any given frequency it will not be as good as a quarter wave despite some of the exaggerated claims made for its performance. A variation on the discone incorporates an additional top element that is often a 5/8th wave whip for a specific band such as the amateur 6m band. Another broadband type for scanner use is what is known as the nest of dipoles. Strictly speaking this is a multiband antenna rather than broadband one, as its peak performance occurs when the dimensions of the elements act as half wave dipoles. This antenna should not be confused with discones that are offered with multiple elements on top. A discone is vertically polarised despite the fact that the top elements are horizontal. Still with multiband antennas are what are known as the sleeved types. These will not usually perform as well as the two antennas mentioned above but are less obtrusive. A good example of this type is the Scanmaster range from Nevada. Finally comes the log periodic. This is a multi-element dipole and for correct use must be operated with an antenna rotator, because it is directional. Commercial ones usually cover 105-1300MHz and 50-1300MHz and offer several dB gain over a discone. The problem of course is that they are directional, but in some cases this can be an advantage if the majority of required signals are being received from approximately the same direction.

Mobile antennas

The true broadband vehicle antenna has yet to be invented. However, compromises are obtained in a number of ways. The first is to use a series of coils along a whip antenna and depending on frequency these can act as resonators or chokes, appearing either high impedance or 'electrically transparent'. The result is a complex multi-band antenna. There are a number of wideband types available from the radio trade, including a glass-mount type which doesn't need a hole drilled or clamps/magnetic mount to be used which could otherwise damage the coax cable as it enters your car via the door seal. The second alternative is to use a nominal 100MHz whip and provide broadband amplification at the base. Finally an interesting solution is an adapter that allows your car radio antenna to feed your scanner as well. The adapter (which is inexpensive and requires no power) is claimed to have no effect on the car radio but there must always be some compromise involved. But then, either you are going to listen to the broadcast radio or you're going to listen to the scanner - surely not both, so a simple coaxial antenna switch, or an antenna relay controlled by the scanner's on/off switch, could be used instead. The final point concerns the scanner antenna plug. It's usually a specific type on car radio leads so if you are using a typical scanner you'll need to change it to a BNC or SMA type.

HF antennas

A common problem found on HF (short wave) by owners of wide-coverage scanners, and even dedicated HF-only receivers, is that when an external 'long wire' antenna is connected the set, rather than hearing a lot more, the end up with a 'mush' of indecipherable signals. This is often the case with sets such as small handheld scanners, which in all fairness are designed to be compact, self-contained sets primarily designed for use with their set-top antenna. Unfortunately, HF reception is often a little better than useless with these, so the natural inclination is to attach a 'purpose-designed' HF antenna, which is when the problems start! So, what can be done?

Switching in the set's built-in attenuator, if fitted, can often help tremendously, although this will also reduce the strength of the wanted signal. For good 'out of band' rejection, some form of extra front-end selectivity is usually needed. You can do this by adding an external manually tuned preselector. This fits in-line between your antenna and receiver,

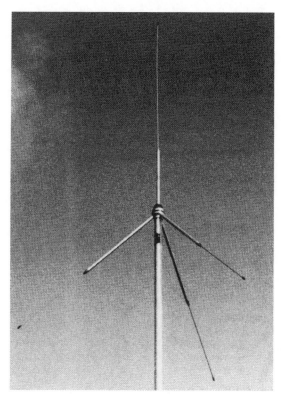

Typical ground plane antenna. This one is designed for airband and the 'drooping' radials give a better 50 Ohm match.

and can sometimes also incorporate a matching unit to allow a high impedance 'long wire' antenna to be connected as well as coax-fed types. There are a number of HF receive preselectors available, from dealers listed at the rear of this book. Adding such a unit will often literally 'transform' your wideband scanner's performance on HF. If you're in the market for an 'off-the-shelf' unit rather than building one yourself, why not take your receiver along to your local ham radio or scanner dealer, and try one of their units on their HF antenna system for yourself before committing yourself? You may be pleasantly surprised, and you'll probably be tempted to come away with a 'new toy'!

The HF 'utility communications' range, typically stretching between around 1.8MHz and 30MHz, covers a number of octaves of frequency, as such a single antenna for the entire range is often a compromise. But the good news is that for receive-only use, a single length of wire will often suffice, in fact the biggest problem you'll get is usually having too much signal level, usually from high-powered broadcast transmitters in adjacent frequency ranges to which you're tuned to.

If your receiver has good front-end selectivity, that is the ability to reject nearby frequency ranges to that which you're tuned to, you'll normally get good results. But often you'll find this only on purpose-designed HF

This discone uses staggered length elements to achieve a wider coverage.

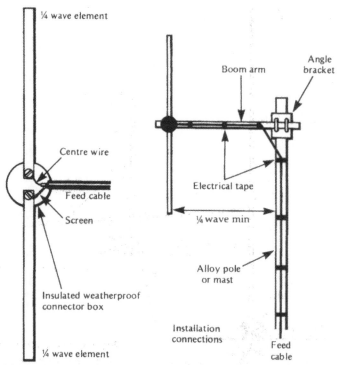

¼ wave element

Boom arm

Angle bracket

Centre wire

Feed cable

Electrical tape

Screen

¼ wave min

Alloy pole
or mast

Insulated weatherproof
connector box

Installation
connections

Feed
cable

¼ wave element

Figure 4.4 The dipole. Vertically polarised, it provides good all round reception.

receivers, which usually means physically 'big' tabletop receivers. A tiny hand-held will often need a bit of help in getting rid of unwanted signals, I'll come onto this a little later.

But let's start with antennas. Because of the longer wavelength of HF frequencies to that of VHF and UHF, your typical VHF/UHF scanner antenna, whether this be a set-top helical, short telescopic whip, or rooftop discone, will be of little use for tuning into relatively low-power (compared with broadcast stations) utility transmitters. You'll instead need to connect something physically longer. Like a length of insulated wire, anywhere from, say, 4m to 40m in length, sited as much 'in the clear' as possible. Even if this means simply throwing a length of wire out of the window, and connecting one end to the centre pin of your receiver's antenna connector. If you can support both ends of the wire away from the ground, for example one end on a washing line post and the other to an upstairs window near to your receiver, all the better. Remember to insulate each end of the wire from its support.

For receive use only, there's no need to buy dedicated antenna insulators, and a handy tip is to use the very strong flexible semi-translucent plastic mouldings you find hold a 'four pack' of cans together. Fold the four sections together to form one circle and you have an instant HF wire

A multiband mobile antenna.

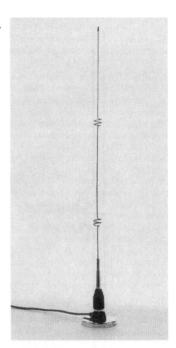

antenna end-insulator! So here, you can get yourself a low-cost start, probably for no more than a few pounds just for the cost of the wire and maybe an antenna connector (I'm not including the price of the four-pack contents!).

Alternatives and Improvements

If your wire is fairly long, you could experience overloading problems, where your receiver just gives you a load of 'mush' from the speaker. Here's where a switchable attenuator in your receiver will help, although this will naturally also cut down the level of the wanted signals. If you can't get an outdoor length of wire, then if you have, say, a loft or outdoor-mounted VHF/UHF antenna, or a TV aerial on your chimney, then just connect both the inner and the outer of the coax (yes, both joined together) to connect to just the inner conductor of your scanner's antenna connector. This will make the length of your coax act as a vertical 'long wire', albeit indoors, but it's better than nothing! Alternatively if you have a washing line with a steel inner core, you could even use this to 'get you going'. I used to use my HF insulated wire occasionally as a washing line, so it can work both ways.

Another improvement, which again will cost very little, is to add an earth connection to your HF antenna system. This means connecting the outer of your receiver's coax socket, using as short and if possible as thick a length of wire as possible, to a 'ground' connection. This ground ideally would be a long earth rod hammered into the ground (watch out for underground water and drainage pipes plus electricity cables!), but it could also be a water pipe, preferably the rising main in your house or the earth connection to the incoming water supply stop-cock at this as often plastic water main piping is used nowadays, or even a nearby central heating radiator to your receiver. *Don't* connect it to a gas pipe! But remember that an electrical earth is usually not very good noise-wise, due to the proximity of the live and neutral wires running in parallel with it around your house.

Matching

If you're serious about HF listening, the next step is to use some form of 'matching' between the long wire and your receiver. As the entire length of the wire will be acting as an antenna, this will include the length running indoors to your receiver as well as the length outside. This indoor length can pick up unwanted signals, e.g. from PCs, fluorescent lights and the often extremely RF-nasty 'energy saving' bulbs which we're now involuntarily being forced to use everywhere, which often have no RF emission filtering built in despite government legislation requirements (low selling price comes first!). To overcome this, a purpose-made long-wire 'balun' (balanced to unbalanced transformer) can be used at the end of the outdoor section of the long wire nearest to your receiver, to transform the wire feed into a coaxial feed, from which you simply run a length of coax to your receiver. I've used a 40m band 'Zepp' antenna for this in the past with good results, alternatively you could use your own length of wire with an 'MLB' - Magnetically Loaded Balun (sometimes described as a Receive Wire Balun) for this.

But one of the most significant improvements you could make to your HF receiving antenna system is to add even a simple Antenna Tuning Unit at the receiver end. As well as matching the long wire or coax input to the impedance of your receiver, it'll add a degree of selectivity. The results can sometimes be little short of outstanding, signals that were previously buried by 'mush' suddenly come into the clear. As well as commercial offerings, ready-to build kits are available. I've also used my own homemade tuning unit, built up from bits from a 'junk box', for many years with great results.

Antennas for portable scanners

A hand-portable scanner can be used with virtually any type of mobile or base antenna and the consequent performance will be better than any small antenna that is normally attached directly to the scanner. However, use of such antennas defeats the portability feature of the scanner. Most

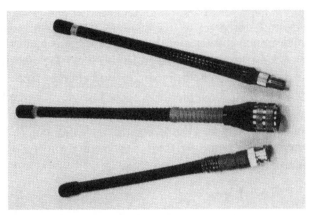

Helicals or 'rubber ducks'. Ideal for handheld scanners, they are available with BNC, PL259 and screw-in connectors.

handheld scanners are supplied with a set-top helical or occasionally a telescopic antenna, and to this end most portable scanners have a BNC type, or more recently a smaller SMA socket, for this which allows the connection of an external antenna. By far the most common antenna used on portable scanners is the helical whip.

Helicals (rubber ducks)

A helical antenna comprises a metal spring, shrouded in a rubber or plastic flexible outer sheath. They are often seen on walkie-talkies and offer the advantage that unlike a telescopic they are flexible and not easily broken. In practice, the spring usually consists of a metal wire wound over a width of about 10mm. The top part is sometimes stretched out slightly but often the lower end is fairly close wound so as to provide correct impedance matching with the scanner. Helicals offer better performance than the loose wire antenna of the same length, are very compact and portable, do not easily break and, being far shorter than respective telescopic antennas, are less likely to do personal damage like poking someone's eye out. However, it is likely that a telescopic antenna will give better performance. Whatever kind of antenna is used, remember that if the scanner is kept in a pocket or any other position close to the body, the performance will be reduced. Not only does the body act as a screen but also the sheer mass can upset the antenna impedance. Portable scanners do not work very well inside vehicles or buildings. In each case it should be possible to attach an external antenna to the set to improve performance. In the case of mobile operation an external magnetically mounted or window mounted antenna can be useful.

Coaxial cable

All base and mobile antennas need to be connected to the scanner using screened coaxial cable. However, any old cable that you have lying around

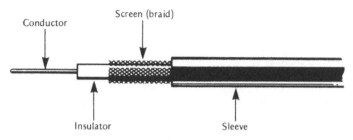

Figure 4.5 Coaxial cable.

might not be suitable. There are several types of coaxial cable available, some for antennas, others for audio and hi-fi use. The latter are unsuitable, as are cables designed for use with a normal car radio. The remaining antenna cables fall into two sections, 50 ohm and 75 ohm impedance types. Cable impedance is important. The 50 ohm variety is normally used to connect professional, commercial, amateur and CB antennas while 75 ohm cable is used for domestic VHF radio as well as terrestrial and satellite television. You must use the right one for your scanner. The scanner instruction manual should tell you which cable impedance to use, and it's invariably 50 ohm. This is available from amateur and CB radio dealers, but like the 75 ohm types two main kinds are available; normal (e.g. RG-58) or low-loss (e.g. RG-8 or one of the many newer types of low loss cable which are increasingly becoming available). If the scanner is only being used for HF and maybe VHF reception and only about 8 or 10 metres of cable is to be used then normal cable can be used. However, if UHF reception or long cable lengths are required then low loss cable is needed – even though this is thicker and less flexible than the lower-cost and higher-loss type.

Cable construction

Whatever kind of coaxial cable you use, it will have roughly the construction shown in Figure 4.5. Starting at the middle is the core wire, which carries the signal, shrouded in a plastic insulator. The insulator prevents the core from touching the outer braid, which is wound in such a way as to provide an earthed screen for the core. This screen serves two purposes: it ensures that the correct impedance is maintained along the entire length of the cable; it stops any interference from reaching the inner core. Finally, the entire cable is covered in plastic insulation. It is important when installing an antenna that this outer insulation is intact, and not torn or gouged so that the braid is exposed. Rainwater getting into the cable in such circumstances will almost certainly ruin the cable. As the screen of the cable is earthed, it is important when connecting the cable to either the antenna or the connector plug that the inner wire and the braid wire never touch. If they do, the incoming signal is earthed, and so lost.

Connectors

In order to plug the antenna cable into the scanner you will need appropriate connectors. If you intend to fit your own then note that you usually will need a soldering iron: twisted wires or wires poked into sockets will almost certainly lead to signal losses. Several types of plug are in common

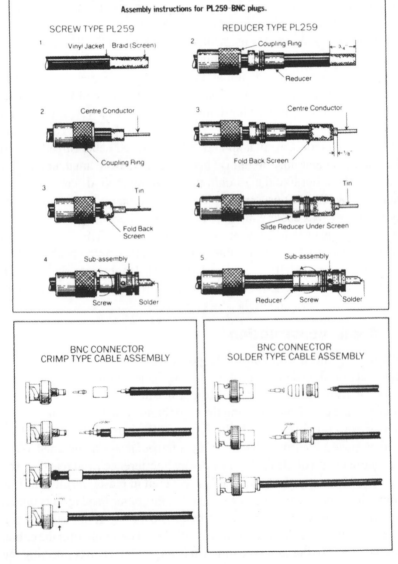

Figure 4.6 Various plugs and connectors (courtesy SSE (UK))

use and are shown in Figure 4.6. Some 'solderless' connectors are also available, and it is also possible to purchase adapters so that an antenna that has one type of plug can be connected to a scanner that takes another type.

PL259; Commonly found on CB sets and amateur HF and VHF equipment, PL259 connectors are only occasionally encountered on scanners, usually of the older variety. However, they are used extensively to connect to antennas. For instance a socket for a PL259 plug will be found on the base of many discones and amplified verticals. The plug comes in several varieties, some of which are easier to fit than others. The simplest are designed for use with the thinner, standard (non-low-loss), cable. The cable is trimmed, about 7-8mm of braid is left and folded back over the outer insulator. The cable is then pushed into the plug and the braiding and insulator screwed into the plug's shell. Once fully home, the centre conductor is then soldered or crimped depending on the plug type. Other types of PL259 have a separate inner sleeve which is either a wide or narrow type depending on the coaxial cable used. This type of plug can be fitted to thicker, low-loss, cables and the appropriate sleeve is purchased separately. When fitting, the braid must be worked back over the sleeve. The inner conductor is then soldered in the normal way. It is a good idea with both types of plug to expose more centre conductor than is needed. The surplus can be snipped off after fitting. PL259 plugs have an outer shell that is internally threaded and when mated with the socket, this shell is screwed up tight to ensure firm contact.

BNC; This is the type most commonly found on scanners, it's much smaller than the PL259 and is the connector often found on professional communication and test equipment. It has a simple twist and pull bayonet action for release, which makes it a lot quicker to change over than the PL259. Unfortunately, its smaller dimensions make it more fiddly to attach to the cable. It is difficult to give specific instructions on fitting as construction differs greatly between makes of plugs.

SMA; This is a miniature connector that has been introduced for use on recent types of portable radio equipment. It's a screw type, almost like a very small PL-259. I'd advise against trying to fit your own coax directly to a mating plug, it's a lot easier to instead use an SMA to BNC adapter which you should find is available from any radio dealer who supplies SMA-equipped receivers to hobbyist users.

Motorola type; This is the car radio antenna type of plug that will be familiar to people who have installed or replaced their car radio at some time. Surprisingly in a way, this low quality type of plug (low quality in terms of performance at VHF and UHF) was found on several very early scanners including the SX200N and some early Bearcats. Types

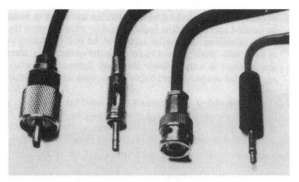

Typical antenna plugs found on scanners. From L. to R: PL259, Motorola, BNC, and miniature jack. The latter is usually confined to handheld sets.

vary with manufacturer, but a look at the plug will usually make it clear how it is attached to the cable.

Miniature jack; A plug that is far from suited to VHF/UHF but it is occasionally found on some of the earlier portable scanners: probably chosen because it is small and very cheap. A look inside will make it obvious as to what is soldered to where, but a difficulty arises in that most miniature jack plugs do not have a big enough opening in the barrel to take 50 Ohm coaxial cable. One way round this problem is to cut off the back end of the barrel with say, a hacksaw, so leaving a bigger opening.

Mounting external antennas

A whole range of fittings and mounting kits is available for installing antennas, and your local TV antenna erection firm should be only too happy to sell you poles, fixing kits, etc., and dealers such as Maplin can also supply a range of fixings. There are four basic ways of mounting an antenna outside and the corresponding kits are:

1 Wall mounting. This consists of a plate with brackets to hold a mounting pole. A hammer drill capable of drilling into brick or masonry will be needed and expanding bolts should be used to retain the plate.

2 Eaves mounting. A smaller version of the wall mounting version, it is used with wood screws to fix onto the eaves. Note, though, that this method is only suitable for small lightweight antennas: even a small antenna can put considerable strain on its mountings during high winds.

3 Chimney lashing. The method often used for TV antennas, comprising one or two brackets held to a chimney by wire cable. Although easy to fit, it places the antenna in close proximity to the heat and smoke from the chimney which may accelerate the inevitable corrosion of the antenna.

4 Free standing mast. This is the most expensive solution but usually the best if you can afford it. An aluminium mast of 6 metres or more in length is partially sunk into the ground and held upright with wire guys. This mounting method can improve performance remarkably at some locations as it allows the antenna to be sited away from obstructions and above the level of trees and buildings that block signals. Planning permission is usually required for this kind of installation.

Antenna amplifiers

Also known as RF or wideband preamps, signal boosters, etc. These devices fall into two categories:

Masthead amplifiers

These units consist of a small-signal amplifier housed in a weatherproof box, physically close to the antenna and usually mounted at or very close to the antenna connection point. They are useful for making-up for the signal losses that occur when long cables are used between the antenna and the scanner. DC voltage to power the unit is fed up the centre core of the coaxial cable - as the signal comes down, the DC goes up, without interference. One variant on this theme is a broadband antenna that actually has an antenna amplifier built into its base.

Cable-end amplifiers

These connect between the end of the cable and the scanner. They are often powered by a small battery, although some are supplied with, or can be used with, a wall plug adapter to plug into the domestic AC supply. Cable-end amplifiers have limitations, as, unlike masthead amplifiers, they cannot improve a poor signal-to-noise ratio of an antenna system.

When to use an amplifier

In the ideal circumstances, that is, with an antenna of sufficient quality and short enough cable, an antenna amplifier is not needed. While capable of boosting weaker signals an antenna amplifier can also cause problems. For instance, strong signals received on other frequencies are also boosted and may overload the scanner. However, some scanner users may live in areas where they are screened by high buildings or land, or may not be able to fit an antenna of sufficiently high quality. In such circumstances an amplifier could help. It may also be of use where long cable runs are necessary between the scanner and the antenna. As a caution, users are advised to seek expert advice before installing an antenna amplifier, as wrongly doing so will cause more problems than it solves.

One antenna, multiple scanners

There are two ways to correctly connect a single antenna to several scanners (don't just wire the coax feeds in parallel!). Each involves the use of a coax 'splitter', and you've the choice of a 'passive' splitter (i.e. one which divides the signal without any amplification and typically with some loss) or an 'active' unit (often called a distribution unit, which amplifies the signal first to overcome the loss of a splitter prior to dividing the now-amplified signal between the outputs).

A passive splitter can be as simple as three resistors connected together in a 'star' formation, although this will lose 6dB (half of the signal), although slightly more expensive hybrid splitters are available that use a combination of coils and capacitors, either are available from TV and DIY stores. If you use one of the hybrid types make sure it covers the frequency range you're after, if it's OK for broadcast FM radio, DAB radio, and TV you should be OK on VHF and UHF, but frequencies below around 75MHz will probably start to be attenuated.

The alternative is a wideband distribution amplifier, again available from TV and DIY stores, but take a look at the technical specifications before you put your hand in your pocket. What you're looking for is a frequency range covering that which you're interested in (fortunately wideband units are usually cheaper than those with a limited frequency coverage!), and without too much gain. In fact the less gain the better, as your scanner will usually already be sufficiently sensitive and any significant amount of additional gain can cause problems with signal overload. Aim for around 6dB maximum and you should be OK, but if it has around 16dB or even 20dB then give it a wide berth unless you happen to live well away from dense areas of radio users! Some commonly available types cover 47 to 862MHz with just a few dB gain which will be fine for VHF and UHF scanner listening.

Blocking and image problems

Whenever you use your scanner, especially with a well-sited antenna connected, you may experience problems from other strong signals in your area. One of these is 'blocking', where the sheer strength of the other signals overloads your receiver and causes the wanted signal to either disappear into the noise or be 'taken over' by the unwanted stronger signal. The other is what's called 'image' reception, where another signal which is a given frequency offset away comes through on the IF 'image' of your scanner. To explain this, if you've a scanner using the common 1st IF of 10.7MHz, then the 'image' frequency of the main ACARS frequency, 131.725MHz, is twice 10.7MHz, i.e. 21.4MHz, above this, on 153.125MHz. Unfortunately this is 'smack bang' in the middle of the VHF paging band, where a strong transmission on this frequency could easily swamp the wanted ACARS signals. To overcome this, you can add a suitable filter in line with the coax antenna connection to your scanner, and Garex Electronics market a useful tuneable type. In the case of VHF airband, you could try an in-line filter such as the AOR ABF125, available from AOR (UK). This is a purpose-designed airband range 'bandpass' filter, which attenuates signals outside of its range. If

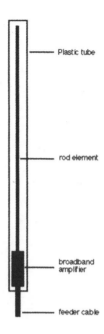

Plastic tube

rod element

broadband
amplifier

feeder cable

Figure 4.7 The active whip for base use.

it's just a specific frequency you want to attenuate then a simple homemade coaxial stub filter is another option. I've used these on numerous occasions in the past, the arrangement being of a coax 'stub' connected in a 'T' formation with the coax down-lead to your receiver. The stub needs to be a precise electrical quarter wavelength at the frequency you want to attenuate, taking into consideration the 'velocity factor' of the coax used for the stub. For both RG213/UR67M (the thick type approx. 10mm diameter, often used for down-leads) and for RG58 & UR76 (the thinner types, around 5mm diameter) the velocity factor will be 0.66. The overall length of the stub in mm will be 75000 divided by the frequency, multiplied overall by the velocity factor. So, in the case of coax types above, the length in mm will be 49500 divided by the frequency in MHz, and at 153.125MHz you'll need a stub length of 323mm. In practice, the velocity factor can vary very slightly between batches of coax, and your 'T' termination will also have some effect, so I'd suggest making the length about 20mm longer and carefully cutting off, say, 2mm at a time for maximum attenuation of the unwanted signal. The end of the stub should be open circuit, i.e., with the inner conductor and outer braid not touching each other. You can either use a BNC 'T' piece with BNC connectors at each coax end, or just solder the inners and outers together directly. At these low VHF frequencies, even one of the small versions of 'chocolate block' connectors could be pressed into service at a pinch.

A group of various scanner antennae on a roof.

Chapter 5
Radio Systems
Explained

After you've 'started out', have you ever wondered why you can sometimes hear part of a conversation, whilst on some systems conversations seem to appear and disappear in mid-conversation. Also, what are those strange tones and other signals heard on the bands? Hopefully this chapter will be of use to explain at least some of what you hear.

Single frequency simplex

This is when the radio communication you're listening to takes place on one frequency, let's say 145.550MHz. Station 'A' has his say, then releases his microphone push-to-talk to let station 'B' reply. Station 'B' uses the same frequency to transmit on as station 'A', and if station 'C' wants to join in, then providing he's in range of both all he has to do is 'butt in' between transmissions. Typical examples are aircraft communication, PMR446 walkie-talkies, and CB operation. This is all nice and simple, so why can't it all be like this. Surely for 'push to talk' communication it would make sense for each user just to use a given frequency for this? In some cases, maybe not.

Dual frequency simplex

This is where one user transmits on one frequency, the other on a different frequency, with their corresponding receiver set accordingly. You'd think that with the shortage of frequencies (many of the bands are getting very congested, and some are literally 'full up' in some cities with no more room for new users) this is rather a waste of a valuable resource - the radio spectrum. But it's not always that simple.

Take the case of a PMR (Private Mobile Radio) service. The base station antenna is often well sited, communicating with mobile and portable stations with their usually lower power transmitters and antennas at ground level. Here, the base station would typically transmit on one frequency, and receive in the same band but on a different frequency, with the mobiles/portables operating on the reverse of these. The communications range of a base station to another base station sharing the same frequency would be fairly large, but not for the mobiles/portables, which often only need to communicate with their 'own' base. So, with a bit of careful planning of frequencies by individual country's licensing bodies, taking hills and so on into account, better frequency re-use can be achieved than with a 'simplex' system.

You may sometimes happen upon a transmission on one 'half' of such a system, apparently communicating with someone else you know you should be able to receive. In this case, you now know that you need to look for the 'other half' on a different frequency.

Talkthrough

But what happens when two mobiles want to communicate with each other? As they transmit and receive on different frequencies, this would be rather difficult without 'external help'. This comes in the form of the base station where it has radio frequency filtering circuits added to allow simultaneous transmission and reception on its separate frequencies. With its receiver audio linked to the transmitter audio, this allows it to relay the signals received on its well-located antenna system, using the higher power base transmitter and the same well-located antenna. This is commonly called, quite simply, a 'repeater'.

This can allow users to communicate even when there's no operator present at the base station desk, for example out of office hours. If a night watchman for example is issued with a portable radio, he can use the repeater to communicate with other employees who are also 'on the radio'. At other times, the base station control operator can switch 'talkthrough' on and off depending on operational requirements, thus explaining why you can sometimes hear both sides of the conversations, and sometimes not. Again, you may happen to stumble on the 'input' frequency of such a system, with users apparently communicating with each other - in this case you'll need to look for the 'output' frequency.

Community repeaters

A well-sited repeater system doesn't need to be restricted to a single user. Many PMR users only require short periods of communication throughout the day, so a better use of such a valuable resource is often

Directional antennae used for the lineside trunked radio system used on the UK's railways.

'shared' amongst several different groups of users, with the 'base station' operator of each simply having a radio transceiver on the same transmit/ receive frequencies as the mobile and portable units, the repeater relaying the audio of all these. To prevent disturbance and maintain a degree of privacy between groups of users, CTCSS (Continuous Tone Controlled Squelch System) and DCS (Digital Code Squelch) is commonly used. Here, the transmitters of each group automatically radiate a low sub-audible CTCSS tone, typically between 67Hz and 250Hz in audio frequency, or a digital code (which shifts the radio frequency back and forth very slightly which is again inaudible) along with their speech, which the repeater regenerates. The accompanying receivers have decoder circuits fitted which only enable the receiver audio when the correct tone or DCS code is present, keeping it quiet otherwise, i.e., when no signal, or a signal with a different tone, is received. To prevent interference between users, each user's transmitter circuit is disabled if the receiver senses an incorrect tone, preventing one user from accidentally transmitting 'on top of' another.

Many such repeater systems are in use, and although a casual monitoring check may reveal several different users, they're all communicating in relative privacy from each other.

Wide Area use

A community repeater is fine for communication in a given area, but what about users who want communication over a larger area, one which can only be given by a number of such base station sites? One way is to have a number of simultaneously interconnected base stations, all operating on the same frequency (these actually operate with a carefully controlled few Hz difference in frequency between them). Here the mobile user's transceiver stays on a given channel, and the base station's messages (and the relayed 'talkthrough' audio from other mobiles/portables) is transmitted over all the base stations simultaneously. The few Hz difference involved in this 'quasi-synchronous' system can sometime be heard as a semi-rapid 'fading' in the received signal if the receiver is stationary - this effect all but disappears when the user becomes 'mobile'.

Trunking

An alternative method, making use of 'intelligent' multi-channel radios operating under microprocessor control, makes use of a technique called 'trunking'. Here, multi-channel two-frequency base stations are sited at strategic points in the required coverage area, each covering an individually defined area 'cell'. They transmit a continuous 'system control' data stream on one of their operating channels, with several other channels at each base station being used as needed for actual communication. The mobile radio automatically searches out this 'system' channel for the area it's in and locks onto it, automatically sending a short 'I'm here' data signal to the base station giving the mobile's identity. When a call comes in, the 'system channel' sends out a data sequence to the mobile, typically instructing it to automatically shift its channel to one of the communication channels in use at that base station, where communication takes place. Each base station is linked to the other, so conversations can take place between sites. As the mobile unit travels around, its channel can be automatically controlled by the base stations to 'move' to that of a neighbouring site, this often happens without the user realising he's being 'handed off' between base station 'cells'.

The use of 'trunking' isn't new; it's been used on landlines and also for STD (Subscriber Trunk Dialling) for many years. Its use in radio systems however is increasing every day, as it's an efficient means of using a limited number of communications channels. Trunking dynamically allocates a radio channel, usually a two-frequency channel with each end transmitting and receiving on different frequencies, and with at least the 'base' end operating in full-duplex mode.

Trunked radio systems operate in a similar way to telephone trunking. Here, a given number of 'lines' are available, between two cities for example, with a much larger total number of 'users' who have access to these lines. Not all users wish to have communication at the same time, so the lines are allocated as and when needed, for the exclusive use of that call but only for as long as it's needed. Radio trunking is basically the same, where a given number of radio channels in each area are shared between a large number of radio users, the radio units having built-in circuitry to automatically change channel as and when needed. In other words, they're 'intelligent' radio units. A 'pool' of radio channels is available which are used as and when needed by each radio, after which the channel is returned to the common 'pool'.

Typical examples a trunked systems can be found in the UK Band III allocation, used for local councils, road and rail transport, and some 'community repeater' systems are operating with multi-channel trunking. Here's how it works.

Multi-channel

For each base station, a number of radio channels are used, normally with the rack of transmitter/receivers combined into a common antenna system or systems. One of the channels for each coverage area is known as a 'control channel', this usually sends out a constant stream of data giving registration and call information. Some PMR systems may 'cycle' this channel between the other channels, or indeed 'time share' it with control channels at other base station sites in the same system.

When a mobile radio unit is switched on, it first hunts and 'locks on' to this control channel. It then checks that it is the correct system etc. that the mobile is registered for, if not then it hunts for another control channel from its pre-stored list of frequencies. Once it's confirmed this, it sends a short burst of 'registration' data to the system, to say 'I'm switched on and available for communication at this site'. This is done on the 'reverse' control channel, i.e., the split-frequency that the radio system is 'listening' on. The trunking system controller updates itself with that information, to the effect that such-and-such a mobile identity is 'logged on' to that base.

Calling

When the radio user places a call, the mobile again transmits a short burst of data to the system, on the 'reverse' control channel. If a speech communication channel is available, the system, again via the control channel, instructs that mobile to automatically shift frequency to the allocated channel, which is used until the end of that period of communication. After the end of the communication, the mobile reverts back to silently monitoring the control channel, and the communication channel which was used is placed back into the 'pool' of available channels for all users.

Cellular coverage

Because of the multi-frequency availability by both base stations and mobile radios, combined with carefully 'tailored' coverage areas for each base station transmitter site using VHF and UHF frequencies, the same frequencies can be re-used a given distance away. By adding coverage areas in this way, a 'network' based on 'cellular' coverage areas can be achieved, with overlaps between 'cells' to ensure that mobiles moving in between these cells can always achieve communication on one channel or another. Neighbouring cell sites must of course use different frequencies for the 'control channel', plus different frequencies for the communication channels, to avoid problems to mobile stations at overlapping signal areas.

Automatic handoff

When a mobile moves from one cell coverage area and into the next, the signal strength from the first control channel, which the mobile is silently monitoring, gets progressively weaker. At a given level, the set's 'brain' decides 'this is too weak, I'll try and find another' and commences to scan the frequencies in its internal memory to try and find a stronger control channel. When it does, it again re-registers with a data burst on the reverse control channel to update the system as to its location. Any incoming calls to that mobile will then be routed to that site rather than the site it was last registered on.

When the mobile radio user is in the middle of a conversation, and starts running out of coverage of the first cell and into that of another, an automatic 'handoff' can occur. Here another data burst automatically instructs the set to shift frequency to that of a given communication channel allocated for its use on the neighbouring site.

Monitoring

When scanning across a given frequency range used by a trunked system in situations where you are allowed to monitor such things, you'll typically hear a channel constantly transmitting a high-pitched 'warble', 24 hours a day. This is the control channel, and if you listen carefully you may hear slight changes, this being the data transmitted as calls are made and the like. A number of 'communication' channels are associated with each control channel, when these are in use at any given time you'll hear normal speech possibly combined with data bursts as the mobiles are automatically instructed to very their status and so on, and finally a burst followed by the signal dropping in mid-conversation when the mobile is either instructed to move frequency onto a channel on the next site or when the call has been completed. The channels used are invariably split-frequency, i.e., mobiles transmit on one frequency and receive on another. Some are half-duplex, i.e., 'push-to-talk' where only hear one side of the conversation is normally heard if communication is being made between different sites.

TETRA

TETRA is TErrestrial Trunked RAdio, which is a digital radio system eminently suitable for emergency services. It gives high speech quality in noisy conditions and full duplex speech, as well as text and even digitised video transmission capabilities. It operates on 380-400MHz across Europe, with the facility for up to 4 communication channels on each radio channel frequency. Radios can also have features such as a panic button and even automatic position finding. Currently virtually all police forces in England, Scotland and Wales use this, as well as the system being available to a significant number of other emergency services. Due to the nature of transmission and inherent digital encryption, it is not possible to monitor TETRA with something like a typical hobbyist handheld scanner. However professional monitoring equipment is available for this, albeit at a professional price!

Speech encryption

Many services naturally want to keep their on-air conversations private, usually not primarily from scanner enthusiasts who just listen in for a hobby but from others who could use the information to their advantage. Typical cases could be a taxi firm with other firms 'poaching' their business, as well as both private and government security services such as store detectives.

Digital communication, using services such as TETRA or GSM / 3G cellphones are one answer of course, and TETRA has a high-level of built-in encryption available for governmental use. Digital scrambling is also possible on normal PMR channels, this is not commonly found in the UK for non-governmental purposes over a wide coverage area, although simple types of scrambling for on-site PMR use, e.g. security guards using portable to portable communication with handhelds, or for hospital and prison safety communication, is often used. Digital scrambling is used in higher-security applications, where the audio is digitised and encrypted before transmission, then digitally unencrypted and converted back to audio at the receiver. The resultant transmitted audio, due to the pseudo-random nature of the digital transmission, often sounds just like white noise on a normal receiver.

Analogue forms of encryption are less secure, but usually have the advantage of being small circuits with much lower current consumption than their digital counterparts, making the overall portable radio smaller and lighter to carry around. 'Add-on' analogue scramblers for retrofitting in normal radios are also available. Some of these use a rolling-code technique, which can be identified by periodic 'bursts' of synchronising data pulses along with the scrambled speech. Others, even simpler, use what's known as frequency inversion, and this system is fitted to several types of widely available PMR handhelds. Here, the user's speech is simply 'inverted', with low frequencies coming out as high frequencies and high frequencies coming out as low frequencies. At the receiving end, the reverse simply happens to result in readable audio again. It's interesting to note that some scanners, such as the Yupiteru MVT9000 and Alnico DJ-X2000E have a switchable voice inverter incorporated, and a base AOR scanner as well as their AR-8200 can have an optional inverter unit fitted.

CTCSS (Sub Tone)

CTCSS stands for Continuous Tone Controlled Squelch System and it's often called 'Sub Tone'. This is because it uses a system where a speech transmission is accompanied by a continuous low-frequency tone, in the range 67Hz to 250.3Hz, i.e. lower than the 'speech range' of 300-3000Hz on a radio system. It's widely used in mobile radio fields, as well as an access method for amateur VHF/UHF repeaters, to give a degree of 'discrimination' between different users on the same channel.

On the air, each user's transmitter generates a given CTCSS frequency, at a low level, along with the speech transmission, and a special decoder is fitted in the receiver to detect only that individual tone. When it does, it opens the receiver squelch, otherwise the speaker remains silent, with the radios of each individual 'fleet' or group of users each being fitted with an identical tone frequency but one that is different from other users. This way, a number of different groups of radio users all on the same channel can use that channel without being overheard by others - that is unless you're listening with an 'open' receiver such as a scanner, when everyone will come through! Often, there's a 'busy channel lockout' fitted in PMR (Private Mobile Radio) transceivers, so that a given user can't transmit if their set is receiving a different tone to 'it's own', to prevent it interfering with other stations. A 'common base station' repeater is often shared by a number of different PMR users, each having a different tone, on the basis of time-sharing a valuable resource such as a hilltop-located repeater station, which automatically re-radiates the signals from mobile and portable users from a well-located site. The repeater re-radiates the CTCSS tone as well as the speech, and often also has 'time-outs' fitted for each user group, to make sure each group gets a 'fair share' of the available airtime.

CTCSS Frequencies

67.0Hz	107.2Hz	167.9Hz
71.9Hz	110.9Hz	173.8Hz
74.4Hz	114.8Hz	179.9Hz
77.0Hz	118.8Hz	186.2Hz
79.7Hz	123.0Hz	192.8Hz
82.5Hz	127.3Hz	203.5Hz
85.4Hz	131.8Hz	210.7Hz
88.5Hz	136.5Hz	218.1Hz
91.5Hz	141.3Hz	225.7Hz
94.8Hz	146.2Hz	233.6Hz
97.4Hz	151.4Hz	241.8Hz
100.0Hz	156.7Hz	250.3Hz
103.5Hz	162.2Hz	

DCS

DCS, which is short for Digital Coded Squelch, acts to the user in a very similar manner to CTCSS. But rather than a continuous sub-audible tone, the actual carrier frequency of the radio signal is digitally 'shifted' in phase by a unique recurring code. Listening on-air to a DCS-coded transmission using a typical receiver is just like listening to a normal transmission, DCS doesn't 'code' or 'encrypt' the signal to prevent others overhearing it. What is does do is to code the signal so that other receivers can be fitted with a DCS decoder, so that their squelch will only open when the correct code is present on the signal. As with CTCSS, this means that many users can effectively 'share' a channel, with relative privacy from each other, as long as different groups don't want to use it at the same time in the same area of course. Some two-way radio users do however seem to be convinced, maybe through enthusiastic and possibly sometimes unwitting sales talk by the seller of the radios, that using DCS makes their communications secure. Maybe so from other two-way radio users sharing the same channel with their radios programmed with a different code. But again as with CTCSS, if you're monitoring with a scanner you'll hear all transmissions on that channel regardless of whether they're using DCS or not.

Selective signalling

In many cases on a radio system, an "all informed net" where all users hear what's happening on the radio channel is very useful. This is typical in an security environment such as a Pubwatch or Shopwatch system where all hearing all communications on the 'local' channel is advantageous, as other officers could either benefit from this or even offer further information from what they hear. But the need to be able to selectively call an individual radio user, or a group of users, has been a desirable ideal in a number of radio systems, for both government and business mobile radio services.

Let's say there's an undercover squad who only want to hear messages intended for them, where a radio bleating out an intruder alarm activation miles away could 'blow a cover'. Also for a council's parks and highways maintenance unit, who'd otherwise be distracted by having to listen out for all messages on the channel while they're working away from the vehicle, just in case one of the radio calls was intended for their unit to respond to?

One method of such individual calling is of course cellular telephones. These of course have an 'airtime charge' of a certain cost per call or time of call, but the very great limitation of these for the 'all informed net', e.g. for emergency services or security officers on a factory site, is that there's no facility whatsoever for 'group' calling. i.e., a call to several people such as 3, 6, or whatever, all at the same time.

Being able to individually call a specific mobile or portable radio unit, or a defined group of these, can this be of benefit. For example, "Any unit able to attend an incident southbound M1 Junction 12" is useful in an 'all-informed net', but if a subsequent call to the specific unit for an update of its arrival at the incident is needed, when the crew would be otherwise engaged in attending to the incident and any other matters arising from it, where 'general' radio traffic might become a background intrusion, a specific call would be beneficial.

5-Tone

Across the PMR communications spectrum, many radio transmissions can be heard being preceded by a short musical tone burst sequence, of around a quarter second to up to a second in duration. This is sequential tone signalling, or it's more common name of "5-tone" from the usual number of individual tones used for each sequence. Here, each mobile radio is allocated a unique 5-digit ID, which a decoder in each radio is equipped to automatically respond to. When the base station dispatcher wishes to call an individual radio user, they enter the radio's ID onto a 5-tone keypad or a PC-based dispatcher-based system, and the transmitter automatically fires up with the tone sequence according to the entered ID.

When the transmitted tone sequence matches the one stored in the receiving radio, it opens the audio path to the speaker so the user can hear the dispatcher's call. Otherwise it remains silent with the user being blissfully unaware of all the other radio activity happening on that channel. Sometimes the mobile radios in a given system are equipped to automatically transmit their own 5-digit ID each time the user presses the radio PTT (push-to-talk), to automatically identify themselves to the dispatcher. A variation to this is a 'status' transmission, where the mobile user can manually change the final digit between 0 and 9, or the final two digits (e.g. 00-99) of their transmitted 5-tone ID, according to their 'status'. So 0 could indicate 'off duty', 1 'on duty', 2 'at customer's premises', 3 'awaiting a new job' and so on, in the case of a two-digit status one of the digits could typically be used indicate the general location area of the mobile at a given time. The first three or four 'fixed' digits of the 5-tone sequence are thus the mobile's individual ID, the final one or two manually variable digits being their status at any time. Usually these status digits are selected by an alphanumeric display on the radio's front panel which indicates the user's status so they can select from a list, rather than having to remember numeric codes.

With such mobile 'status' facilities, the dispatcher often has the facility of automatically 'interrogating' the mobile with a 5-tone call to that mobile, where on receipt it will automatically transpond with it's ID and the status that's been set by the user. This way the dispatcher can see if a radio user is available for a job or not, without the need for a voice call.

It can also work the other way, where in a busy radio system all that a user needs to do is select their status and press a button on the radio to manually transmit this. So, when 'signing on' in the morning or after a lunch break, the radio user only needs to transmit a short burst of 5-tone rather than a longer speech transmission to say they're in the vehicle and either available for work (e.g. in the case of a taxi) or on their way to their first job.

Another use is that of a 'panic' facility, where each radio is also programmed with a unique 5-tone emergency ID that's transmitted if the vehicle's driver hits a 'panic' button on the dashboard. This automatically transmits the mobile's emergency status, possibly followed by a 10 or 20 seconds of 'live microphone' so the driver can call for help over the radio system without the need to fumble about with radio controls.

But in some instances, an ANI is used to prevent radio users misusing their system, e.g. by playing music etc. over the air. Yes, it does happen, and in some countries government users such as the police are the worst offenders! These users with 'trailing' ANI have sometimes learned they can transmit whatever music or profanity they want, then physically switch the radio off before releasing the PTT to prevent the ANI being transmitted. Here, 'radon' ANI is used, when the short ANI sequence is randomly sent at any time during the user's transmission, i.e. they don't know when it'll be sent so they won't risk misusing the radio. If the actual transmission length is shorter than the time when the random ANI would have been sent, it's automatically sent at the end of the PTT press.

The table below shows the most commonly used tone sequences and tone lengths, in case you fancy using PC-based soundcard software to decode these. A 'repeat' (r) tone is used in instances where the same digit is sent for successive tones, to make sure each actual tone frequency is different from the one immediately before it. For example, '12334' would be sent using the tones corresponding to 123R4 (1, 2, 3, repeat, 4)

5-Tone sequences commonly used in the UK and Europe

Tone No.	EEA	DZVEI	ZVEI	ITU-R	DDZVEI
1	1124	970	1060	1124	2200
2	1197	1060	1160	1197	1060
3	1275	1160	1270	1275	1160
4	1358	1270	1400	1358	1270
5	1446	1400	1530	1446	1400
6	1540	1530	1670	1540	1530
7	1647	1670	1830	1640	1670
8	1747	1830	2000	1747	1830
9	1860	2000	2200	1860	2000
0	1981	2200	2400	1981	2200
Repeat (R)	2110	2400	2600	2110	970
Group(G)	1055	885	2800	2400	885
Alarm(A)	2400				
Tone Length 40mS	70mS	70mS	100mS	70mS	

Data bursts

A growing number of PMR (Private Mobile Radio) channels are, besides using speech communication, having the beginning or end of the speech transmissions accompanied by a short data burst. One such use is that of a trunked system, where channels are usually dynamically allocated and the data bursts are used as individual identifications for 'PTT On', PTT Off' as well as an individual ID. But an increasing use is that for an additional data message to accompany the transmission to give status and location information. With GPS (Global Positioning System) equipment becoming cheaper day by day, many users are adding these to their radio system to keep track of vehicles and even portable units.

In use, each radio is linked to a GPS receiver, which can either be built-in or be an external unit that's plugged into an accessory socket on the radio. Each time the radio user presses their radio's PTT (Push To Talk) button, as well as their speech being transmitted the GPS information, which can also include speed, direction of travel, and status information such as engine on, door open etc., as well as the actual location, is also transmitted in a data format, usually as a short 'burst' either at the beginning of the 'over' or as a trailing burst at the end of each transmission. At the base station end, a data modem connected to the base radio, again either built in or as an external unit, is linked to a PC running appropriate mapping and over-air information processing software.

A typical PC display would be an on-screen map of the area in which the mobile fleet is operating in, which the operator can zoom in or zoom out of as needed to focus on a particular area or vehicle. The actual location of each vehicle in the mobile fleet is graphically shown on the map, usually as a small symbol such as a car or whatever along with their numerical ID or a short alphanumeric 'tag' identification. On a separate part of the display is more information about what the vehicle is doing, for example whether it's moving and if so what the speed and direction is, along with any status indications such as engine on, door(s) open etc.

Besides having an update of each vehicle whenever the mobile user transmits a message, each mobile can also be set to automatically transmit this information periodically, every 10 minutes for example, or in some cases whenever it's moved in position by a given distance. Hence if you listen on a channel you'll typically also hear what sounds like seemingly random bursts of data. As well as this, the base station can 'poll' either individual mobiles, or a group of mobiles, to get an instant update on where they are and what they're doing at any given instant. FFSK (Fast Frequency Shift Keying) is invariably used for the over-air data, and in, say, a 1200 baud system this can be one cycle of a 1200Hz tone to represent a data 'zero' and one and a half cycles of an 1800Hz tone to represent a data 'one'.

The question you'll invariably be asking now is "Can I use my scanner to receive this and have a similar display on my PC?" The answer here can be yes, no, or 'maybe' depending on what's being used. Some commercial systems naturally need to be kept reasonably secure, for example in the case of a security company's vans carrying lots of valuable cargo, they wouldn't really like potential criminals doing this as well so that they can stage a hold-up! However, with many systems it's quite possible to link your scanner's audio up to your PC, typically using the PC's sound card as an audio input, and running suitable data decoding software. Here you'll usually get a 'raw' data output display in text format, with information such as vehicle ID, latitude and longitude, and so on, although this won't be in a pretty map just like the base station operator sees. However 'cutting and pasting' into a mapping software such as Google Maps or Google Earth will instantly let you see where that particular mobile is.

However, there is another system that's commonly used, and that's APRS, which stands for Automatic Position Reporting System. It's a packet radio system which was pioneered by radio amateurs but now used by a wide group of radio users, there's freeware and shareware software readily available for this, along with plenty of ready-made maps to use with the software to display the locations and so on of the radio users. Commonly used programs are WinAPRS and UI-View. As well has being able to see the locations of radio amateurs, other users including the Space Shuttle when it's launched also carry APRS equipment, even the International Space Station has a packet station on board.

DTMF

Some PMR equipment uses DTMF signalling (Dual Tone Medium Frequency, as used for telephone dialling) rather than single-tone frequency signalling. This typically uses a string of three digits for each radio ID, although radio-to-radio selective calling can use six digits, the first three being the called party and the last three the calling party, to display on the called radio who's calling. Each digit is made up of an audio tone pair, and there are plenty of decoders readily available to decode and display these sequences, as well as plenty of shareware PC programs that use your sound card as the audio to PC interface.

DTMF Tones

	1209Hz	1336Hz	1477Hz
697Hz	1	2	3
770Hz	4	5	6
852Hz	7	8	9
941Hz	*	0	#

ACARS

If you're an airband enthusiast, then you may have head of ACARS, the Aircraft Control and Reporting System. This is an automatic data system used by civil airlines around the world, to transmit real-time data of flight status reports together with data and text messages involving both passengers and crew. Time of "Wheels-up", touchdown, plus ground and air speed are typical transmissions, along with text-based information such as schedules, weather information, 'status' messages from the aircraft such as engine performance, fuel usage, emergency conditions and even exact times of takeoff, passenger door openings and so on. Other messages can include private or 'company' messages' either between airline staff or for passengers on the craft.

To decode the data, all you need is a suitable interface and decoding program, of which there are a number commercially available such as the Lowe 'Airmaster'. Alternatively, you can use the sound card of your PC as the data interface, running a program such as WACARS, i.e. Windows ACARS. The main frequency used in Europe is 131.725MHz, where you'll often hear periodic data bursts from a number of aircraft. AOR also produce a portable self-contained ACARS decoder for out-and-about use.

An ACARS transmission sounds like a short high-frequency (2400Hz) audio burst, followed by a data stream which sounds very much like an amateur packet radio transmission, i.e. a 'raspy' noise lasting between a half second and several seconds, depending upon the length of the data being sent.

Paging

Personal digital and alphanumeric pagers are widely used, and many listeners will have heard the 'Brrrrrrr...brrppppp' paging signals on VHF and UHF (see the 'allocations' section in this book), which are high powered paging transmitters sending text messages out to predefined users.

The system most widely used for this is called POCSAG, which stands for 'Post Office Code Standardisation Advisory Group', another is a proprietary system called FLEX. But there are very many easy, and freely available, ways of decoding these messages. You'll just appropriate freeware or shareware software running on a PC, using the PC's sound card as the interface.

As POCSAG and FLEX are transmitted as direct FSK (frequency shift keying), the best way is to take the data direct from the receiver's internal discriminator circuit, rather than from the earphone or external speaker connector, although some base scanners like the Icom IC-PCR1000 and IC-PCR1500 models have a dedicated 'data' output facility.

Incidentally, because VHF and UHF pagers use small, internal antennas, the paging transmitters are usually very high-powered affairs in order to give 'blanket' coverage. Unfortunately, because of the proximity of the paging allocations to other frequencies, scanner users often have problems (sometimes extremely severe) with interference from the unwanted pager base transmissions. This is where the technical results in the 'Scanners and Accessories Review' section of this book on *real* scanner performance becomes invaluable, where you can see which sets have better 'blocking' rejection than others before you go out and spend your hard-earned cash!

Chapter 6
UK Frequency
Allocations

The decision on who transmits what, on which frequency, is made by international agreement. Clearly, governments must agree on allocations if they are to avoid causing interference. There would be chaos if, say, one country allocated a band to low powered radio telephones while a neighbouring country allocated the same brand for high powered broadcasting. The body which co-ordinates radio frequency allocations on behalf of world governments is the International Telecommunications Union, known simply as the ITU. For the purpose of agreed allocations the ITU splits the world into three regions. The United Kingdom falls in Region 1, which includes most of Europe and a small section of North Africa. However, it does not necessarily follow that each country conforms strictly to the allocations drawn up for that region. Where there is little likelihood of interference, countries may opt for local variations and, obviously, many such variations exist. For this reason, listings given in this book strictly apply only to the United Kingdom, although most allocations do match the standard format for Region 1. We shall look first at HF allocations, some spot allocations used here, then general VHF/UHF Frequency allocations, then consider in detail some of the services on the VHF / UHF allocations.

HF Bandplans

As many scanners cover the HF (Short Wave) spectrum as well as VHF/UHF, below is a HF bandplan to provide a guide to international allocations. Region 1 (UK and Europe) allocations are given up to 7300 kHz, with common worldwide allocations shown above 7300 kHz.

From	To	Allocation
1605	1625	Maritime Mobile, Fixed, Land Mobile
1625	1635	Radiolocation
1635	1800	Maritime Mobile, Fixed, Land Mobile
1800	1810	Radiolocation
1810	1850	Amateur
1850	2025	Fixed, Mobile (not Aero)
2025	2045	Fixed, Mobile (not Aero), Meteorological Aids
2045	2160	Maritime Mobile, Fixed, Land Mobile
2160	2170	Radiolocation
2170	2173.5	Maritime Mobile
2173.5	2190.5	Mobile (Distress)
2190.5	2194	Maritime Mobile
2194	2300	Fixed, Mobile (not Aero)
2300	2498	Fixed, Mobile (not Aero), Broadcasting
2498	2502	Standard Frequency & Time Sig (2500 kHz), Space Research
2502	2625	Fixed, Mobile (not Aero)
2625	2650	Maritime Mobile, Maritime, Radionavigation
2650	2850	Fixed, Mobile (not Aero)
2850	3155	Aeronautical Mobile
3155	3200	Fixed, Mobile (not Aero)
3200	3230	Fixed, Mobile (not Aero), Broadcasting
3230	3400	Fixed, Mobile (not Aero), Broadcasting
3400	3500	Aeronautical Mobile
3500	3800	Amateur, Fixed, Mobile (not Aero)
3800	3900	Fixed, Mobile (not Aero), Land Mobile
3900	3950	Aeronautical Mobile
3950	4000	Fixed, Broadcasting
4000	4063	Fixed, Maritime Mobile
4063	4438	Maritime Mobile
4438	4650	Fixed, Mobile (not Aero)
4650	4750	Aeronautical Mobile
4750	4850	Fixed, Aeronautical Mobile, Land Mobile, Broadcasting
4850	4995	Fixed, Land Mobile, Broadcasting
4995	5005	Standard Frequency & Time Sig (5000 kHz)
5005	5060	Fixed, Broadcasting
5060	5250	Fixed, Mobile (not Aero)
5250	5450	Fixed, Mobile (not Aero)
5450	5480	Fixed, Aeronautical Mobile, Land Mobile
5480	5730	Aeronautical Mobile
5730	5950	Fixed, Land Mobile

From	To	Allocation
5950	6200	Broadcasting
6200	6525	Maritime Mobile
6525	6765	Aeronautical Mobile
6765	7000	Fixed, Land Mobile
7000	7200	Amateur, Amateur-Satellite
7200	7300	Broadcasting
7300	8100	Fixed & Land Mobile
8100	8195	Fixed & Maritime Mobile
8195	8815	Maritime Mobile
8815	9040	Aeronautical Mobile
9040	9500	Fixed
9500	9900	Broadcasting
9900	9995	Fixed
9995	10005	Standard Frequency & Time Sig (10000 kHz)
10005	10100	Aeronautical Mobile
10100	10150	Fixed & Amateur
10150	11175	Fixed & Mobile (not Aero)
11175	11400	Aeronautical Mobile
11400	11650	Fixed
11650	12050	Broadcasting
12050	12230	Fixed
12230	13200	Maritime Mobile
13200	13360	Aeronautical Mobile
13360	13410	Fixed & Radio Astronomy
13410	13600	Fixed & Mobile (not Aero)
13600	13800	Broadcasting
13800	14000	Fixed, Mobile (not Aero)
14000	14250	Amateur, Amateur-Satellite
14250	14350	Amateur
14350	14990	Fixed, Mobile (not Aero)
14990	15010	Standard Frequency & Time Sig (15000 kHz)
15010	15100	Aeronautical Mobile
15100	15600	Broadcasting
15600	16360	Fixed
16360	17410	Maritime Mobile
17410	17550	Fixed
17550	17900	Broadcasting
17900	18030	Aeronautical Mobile
18030	18068	Fixed & Space Research
18068	18168	Amateur & Amateur-Satellite
18168	18780	Fixed
18780	18900	Maritime Mobile
18900	19680	Fixed
19680	19800	Maritime Mobile
19800	19990	Fixed
19990	20010	Standard Frequency & Time Sig (20000 kHz), Space Research
20010	21000	Fixed & Mobile
21000	21450	Amateur & Amateur-Satellite
21450	21850	Broadcasting
21850	21870	Fixed
21870	21924	Aeronautical Fixed
21924	22000	Aeronautical Mobile

From	To	Allocation
22000	22855	Maritime Mobile
22855	23000	Fixed
23000	23200	Fixed & Mobile (not Aero)
23200	23350	Aeronautical Fixed & Mobile
23350	24000	Fixed & Mobile (not Aero)
24000	24890	Fixed & Land Mobile
24890	24990	Amateur & Amateur-Satellite
24990	25010	Standard Frequency Time Sig (25000 kHz), Space Research
25010	25070	Fixed & Mobile (not Aero)
25070	25210	Maritime Mobile
25210	25550	Fixed & Mobile (not Aero)

From	To	Allocation
25550	25670	Radio Astronomy
25670	26100	Broadcasting
26100	26175	Maritime Mobile
26175	27500	Fixed & Mobile (not Aero)
27500	28000	Meteorological Aids, Fixed & Mobile (CB)
28000	29700	Amateur & Amateur-Satellite
29700	30005	Fixed & Mobile
30005	30010	Space Operation (Satellite Identification), Fixed, Mobile & Space Research
30010	37500	Fixed & Mobile

HF Communications Spot Frequencies (all Upper Sideband unless stated)

Frequency MHz	User
3.081	NATO E-3 AWACS Magic; NB
3.160	Army 352nd SOG in Bosnia Black Hat and Promenade
3.178	Bookshelf Net Push Unknown
3.225	NATO E-3 AWACS Magic; NC
3.357	US Navy
3.900	NATO E-3 AWACS Magic
4.542	NATO E-3 AWACS Magic; NE
4.612	METO-IFOR in Bosnia
4.720	NATO E-3 AWACS Magic; NF
4.724	USAF Global HF System Primary
4.758	NATO E-3 AWACS Magic; NH
4.7775	NATO
5.218	Bookshelf Net
5.245	Air Military
5.267	NATO
5.349	NATO Training NET; Ch661G
5.691	NATO E-3 AWACS Magic
6.693	USN Voice Co-ordination Net
6.693	Military
6.695	NATO E-3 AWACS Magic; KF
6.697	Navy High Command Voice Frequency
6.712	France Air Force
6.728	NATO AWACS Net
6.729	Military Night Frequency.
6.734	Serb Military
6.738	Military Night Frequency
6.739	Air Mobile Command (Air Force)
6.754	NATO E-3 AWACS Magic; XC
6.760	NATO E-3 AWACS Magic; KD
6.761	Military Night Frequency
6.761	USAF Air Refuelling Primary
6.7625	NATO E-3 AWACS Magic; NI
6.773	Military
6.819	METO-IFOR in Bosnia
6.865	Military
6.865	Bookshelf Net Push 81B
6.870	METO-IFOR in Bosnia
6.912	French Air Force
6.9325	Bookshelf Net Push 81B

Frequency MHz	User
7.919	METO-IFOR in Bosnia
8.046	Bookshelf Net Push 81V
8.048	Military USB
8.080	US Navy
8.968	Air Mobile Command (Air Force)
8.971	NATO E-3 AWACS Magic; AB
8.980	NATO E-3 AWACS Magic; A4
8.980	NATO E-3 AWACS Magic; XD
8.9865	NATO E-3 AWACS Magic; NJ
8.992	Air Mobile Command (Air Force)
8.992	US Military GHFS
8.992	USAF Global HF System Primary
9.016	Air Mobile Command (Air Force)
9.023	Air Mobile Command (Air Force)
9.037	Air Mobile Command (Air Force)
9.1185	Bookshelf Net Push 82B
9.260	Bookshelf Net Push 82B
10.315	NATO Naval Net
10.315	NATO E-3 AWACS Magic; A9
10.315	NATO E-3 AWACS Magic; XE
10.780	Air Mobile Command (Air Force)
10.865	US Navy
11.167	Airlift Command Transport
11.173	Bookshelf Net Push 83A
11.175	NATO Standard Frequency USB
11.176	Air Mobile Command (Air Force)
11.228	NATO E-3 AWACS Magic; AA
11.244	Air Mobile Command (Air Force)
11.267	Navy High Command Voice
11.2705	NATO E-3 AWACS Magic; NK
11.271	Air Mobile Command (Air Force)
11.300	Serbia Fighters
13.200	Air Mobile Command (Air Force)
13.200	USAF Global HF System Primary
15.016	USAF Global HF System Primary
15.050	NATO E-3 AWACS Magic; NL
15.959	US Navy
17.967	Air Mobile Command (Air Force)
17.9965	NATO E-3 AWACS Magic; NM
20.015	US Navy
23.214	NATO E-3 AWACS Magic

HF Broadcast stations (AM)

Frequency MHz	Broadcaster
6.115	Radio Tirana
7.115	Radio Yugoslavia
7.115	Radio Sweden
7.125	Voice of Russia
7.130	Radio Yugoslavia
7.160	Radio Tirana
7.180	Voice of Russia
7.195	Radio Bulgaria
7.250	Voice of Russia
9.570	Radio Bulgaria

Frequency MHz	Broadcaster
9.650	Radio Tirana
9.925	Radio Croatia
11.830	Radio Bulgaria
12.000	Voice of Russia
12.020	Voice of Russia
13.820	Radio Croatia
15.595	Voice of Russia

UK Frequency Allocation Table, 30-1525MHz

The following table shows the official UK allocations by the government for radio services and users. Primary users of each segment are given in capitals, secondary services are given in lower case.

From-to pairing	Allocation
30.005-30.01 MHz	SPACE OPERATION (satellite identification) FIXED MOBILE SPACE RESEARCH
30.01-37.5 MHz	FIXED MOBILE
37.5-38.25 MHz	MOBILE FIXED Radio Astronomy
37.5 – 38.25MHz	FIXED MOBILE Radio Astronomy
38.25 – 39.986MHz	FIXED MOBILE
39.986 – 40.02MHz	FIXED MOBILE Space Research
40.02 – 40.98MHz	FIXED MOBILE
40.98-41.015 MHz	MOBILE Space Research
40.015-44.00 MHz	MOBILE
44.0-46.4 MHz	MOBILE
46.4-47.0 MHz	MOBILE
40.98 – 41.015MHz	FIXED MOBILE Space Research
41.015 – 44MHz	FIXED MOBILE
44 – 47MHz	FIXED MOBILE
47.0-50.0 MHz	LAND MOBILE
50.0-51.0 MHz	AMATEUR
51.0-52.0 MHz	LAND MOBILE Amateur
52.0-68.0 MHz	LAND MOBILE
68.0-70.5 MHz	MOBILE except aeronautical mobile Amateur
70.5-71.5 MHz	LAND MOBILE
71.5-72.8 MHz	LAND MOBILE

From-to pairing	Allocation
72.8-74.8 MHz	LAND MOBILE
74.8-75.2 MHz	AERONAUTICAL RADIONAVIGATION
75.2-76.7 MHz	LAND MOBILE
76.7-78.0 MHz	FIXED MOBILE
78.0-80.0 MHz	LAND MOBILE
75.2 – 87.5MHz	FIXED MOBILE except aeronautical mobile
80.0-87.5 MHz	FIXED LAND MOBILE Radio Astronomy
87.5-108.0 MHz	BROADCASTING
108.0-117.975 MHz	AERONAUTICAL RADIONAVIGATION
117.975-137.0 MHz	AERONAUTICAL MOBILE
137.0-137.025 MHz	METEOROLOGICAL SATELLITE (space to Earth) MOBILE MOBILE SATELLITE Space operation (space to Earth) Space Research (space to Earth)
137.025-137.175MHz	METEOROLOGICAL SATELLITE (space to Earth) MOBILE Mobile Satellite (space to Earth) Space operation (space to Earth) Space Research (space to Earth)
137.175-137.825 MHz	METEOROLOGICAL SATELLITE (space to Earth) MOBILE MOBILE SATELLITE Space Operation (space to Earth) Space Research (space to Earth)

From-to pairing	Allocation	From-to pairing	Allocation
137 825-138 MHz	METEOROLOGICAL SATELLITE (space to Earth) Mobile Satellite 5.208A, Mobile except aeronautical mobile (R) Space Operation (space to Earth) Space Research (space to Earth)	235.0-328.6 MHz	FIXED MOBILE Radiolocation Radio Astronomy Mobile Satellite
138.0-141.9 MHz	LAND MOBILE Space Research (space to Earth)	328.6-335.4 MHz	AERONAUTICAL RADIONAVIGATION
		335.4-399.9 MHz	FIXED MOBILE Mobile-Satellite
141 9-143 0 MHz	LAND MOBILE Space Research (space to Earth)	399.9-400.05 MHz	MOBILE-SATELLITE (Earth-to-space) RADIONAVIGATION-SATELLITE
143 0-144 0 MHz	LAND MOBILE	400.05-400.15 MHz	STANDARD FREQUENCY AND TIME SIGNAL SATELLITE (400·1 MHz)
144 0-146 0 MHz	AMATEUR AMATEUR-SATELLITE		
146 0-149 0 MHz	FIXED MOBILE except aeronautical mobile	400.15-401.0 MHz	METEOROLOGICAL SATELLITE MOBILE-SATELLITE METEOROLOGICAL AIDS SPACE RESEARCH Space Operations
149 0-149 9 MHz	MOBILE except aeronautical mobile MOBILE-SATELLITE (Earth-to-space)		
149 9-150.05 MHz	MOBILE-SATELLITE (Earth-to-space) RADIONAVIGATION-SATELLITE	401.0-406.0 MHz	METEOROLOGICAL AIDS SPACE OPERATION FIXED MOBILE Except Aeronautical Meteorological-Satellite (Earth to space)
150.05-152.0 MHz	FIXED MOBILE except aeronautical mobile RADIO ASTRONOMY		
152 0-153 0 MHz	MOBILE except aeronautical mobile	406.0-406.1 MHz	MOBILE-SATELLITE (Earth to space)
153 0-153 5 MHz	MOBILE except aeronautical mobile Meteorological Aids	406.1-410.0 MHz	FIXED MOBILE RADIO ASTRONOMY Radiolocation
153 5-154 0 MHz	MOBILE except aeronautical mobile Meteorological Aids	410.0-420.0 MHz	FIXED MOBILE Space Research
154.0-156.4875 MHz	FIXED MOBILE except aeronautical mobile	420-430 MHz	FIXED MOBILE RADIOLOCATION
156.4875-156.5625	MARITIME MOBILE (distress and calling via DSC)	430-440 MHz	FIXED MOBILE RADIOLOCATION Amateur-Satellite Amateur
156.5625-156.7625	FIXED MOBILE except aeronautical mobile	440-450 MHz	FIXED MOBILE RADIOLOCATION
156.7625-156.8375	Maritime mobile (distress and calling)	450.0-470.0 MHz	FIXED MOBILE
156.8375-174 MHz	FIXED MOBILE except aeronautical mobile	470-790 MHz	AERONAUTICAL RADIONAVIGATION BROADCASTING Mobile
174.0-217.5MHz	BROADCASTING MOBILE except aeronautical mobile	790 – 862MHz	FIXED BROADCASTING MOBILE except aeronautical mobile
217.5-230.0 MHz	BROADCASTING Mobile		
230.0 – 235.0 MHz	FIXED MOBILE Radiolocation		

From-to pairing	Allocation
862-870 MHz	MOBILE
870-876 MHz	MOBILE
876-880 MHz	MOBILE
880-915 MHz	FIXED
	MOBILE
	Radiolocation
915-921 MHz	MOBILE
	Radiolocation
921-925 MHz	MOBILE
	Radiolocation
925-960 MHz	FIXED
	MOBILE
	Radiolocation
960-1164 MHz	AERONAUTICAL
	RADIONAVIGATION
	AERONAUTICAL
	MOBILE
1164-1215 MHz	AERONAUTICAL
	RADIONAVIGATION
	5.328
	RADIONAVIGATION-
	SATELLITE (space-to-
	Earth)
	(space-to-space)
1215-1240 MHz	EARTH
	EXPLORATION-
	SATELLITE (active)
	RADIOLOCATION
	RADIONAVIGATION
	RADIONAVIGATION-
	SATELLITE
	(space to Earth)
	(space to space)
	SPACE RESEARCH
	(active)
1240-1260 MHz	EARTH
	EXPLORATION-
	SATELLITE (active)
	RADIOLOCATION
	RADIONAVIGATION
	RADIONAVIGATION-
	SATELLITE
	(space to Earth)
	(space to space)
	SPACE RESEARCH
	(active)
	Amateur
1260-1300 MHz	EARTH
	EXPLORATION-
	SATELLITE (active)

From-to pairing	Allocation
	RADIOLOCATION
	RADIONAVIGATION
	RADIONAVIGATION-
	SATELLITE (space to
	Earth) (space to
space)	
1300-1350 MHz	AERONAUTICAL
	RADIONAVIGATION
	RADIOLOCATION
	RADIONAVIGATION
	RADIONAVIGATION-
	SATELLITE
	(Earth to space)
	Amateur
	SPACE RESEARCH
	(active)
	Amateur
	Amateur-satellite
(Earth	to space)
1350-1375 MHz	FIXED
	MOBILE (except
	aeronautical mobile)
	Radiolocation
1375-1400 MHz	FIXED
	MOBILE (except
	aeronautical mobile)
1400-1427 MHz	EARTHEXPLORATION
	–SATELLITE (passive)
	RADIO ASTRONOMY
	SPACE RESEARCH
	(passive)
1427-1429 MHz	SPACE OPERATION
	(Earth to space)
	FIXED
	MOBILE
1429-1452 MHz	FIXED
	MOBILE
1452-1492 MHz	FIXED
	MOBILE (except
	aeronautical mobile)
	BROADCASTING
	BROADCASTING-
	SATELLITE
1492-1518 MHz	FIXED
	MOBILE (except
	aeronautical mobile)
1518-1525 MHz	FIXED
	MOBILE (except
	aeronautical mobile)
	MOBILE-SATELLITE
	(space-to-Earth)

Current United Kingdom Frequency Usage, spot Frequencies and bands

From-to Pairing	Allocation
25.0050-25.0100	Standard Frequency, time signals, space research
25.0100-25.0700	Fixed (PTO & Government), Maritime & Land Mobile

From-to Pairing	Allocation
	(Government)
25.0700-25.2100	Maritime Mobile (mostly USB and RTTY)
25.0539-26.1444	Ship-to-shore (SSB)
25.0601-26.1506	Ship-to-shore (SSB)
25.0710	Marine Calling channel 'A'
25.0730	Marine Calling channel 'B'

71

From-to Pairing Allocation

25.0750	Marine Calling channel 'C'
25.0763-25.0898	Marine channels spaced 0.5kHz
25.2100-25.6000	Fixed (PTO & Government), Maritime & Land Mobile (government)
25.2100-25.5350	World-wide coastal stations
25.5500-25.6000	Radio Astronomy
25.6000-26.1000	Broadcasting (AM) plus Radio Astronomy
26.1000-27.5000	Fixed (PTO & Government), Land Mobile (including CEPT CB system), pagers, ISM, Maritime Mobile, model control
26.1000-26.1750	MOD tactical land mobile
26.1444-25.0539	Shore-to-ship (SSB)
26.1506-25.0601	Shore-to-ship (SSB)
26.2375-26.8655	One-way paging systems (new band)
26.9780-27.2620	One-way paging systems (old band)
26.9570-27.2830	Industrial, scientific and medical
26.9600-27.2800	Model control (AM & FM & 1.5Watt maximum power) & Data Buoys
26.9650-27.4050	CEPT Citizens Band radio (mostly NFM but some AM & SSB)
27.4500	Emergency alarm systems for the elderly or infirmed
27.5000-27.6000	Land Mobile (government) Meteorological aids (Sondes, etc)
27.6000-28.0000	Land Mobile (UK CB system) & Meteorological Aids
28.0000-29.7000	Amateur Radio (10 Metre Band) including Russian RS-series satellites, also licence-exempt short range devices
29.7000-30.7000	Space (satellite identification), Mobile (Government) Fixed
29.7000-29.9700	Military 25kHz channel spacing simplex communications
29.7000-30.0100	Satellite identification
30.0250-30.7000	USAF (Europe) mobile communications 25kHz channels
30.4500	US Military 'MARS' radio integration network
30.700-34.5000	Fixed, Mobile & Paging systems (at the peak of

From-to Pairing Allocation

	11 year sunspot cycles many US services operating in this band can be heard)
30.7000 -34.5000	Use by USAF in UK/ Europe and MOD. Mostly NFM at 25kHz channel spacing
31.0375/39.9375	CT1 cordless phones base unit TX
31.0625/39.9625	CT1 cordless phones base unit TX
31.0875/39.9875	CT1 cordless phones base unit TX
31.1125/40.0125	CT1 cordless phones base unit TX
31.1375/40.0375	CT1 cordless phones base unit TX
31.1625/40.0625	CT1 cordless phones base unit TX
31.1875/40.0875	CT1 cordless phones base unit TX
31.2125/40.1125	CT1 cordless phones base unit TX
31.7250	Hospital paging systems
31.7500	Hospital paging systems
31.7750	Hospital paging systems
31.8000 -34.9000	Military Fixed/Mobile 50kHz channel spacing
34.5000-37.5000	Mobile (mostly government & military), Model Control, Alarms
34.9250	Emergency alarm systems for the aged and infirmed
34.9500	Emergency alarm systems for the aged and infirmed
34.9750	Emergency alarm systems for the aged and infirmed
35.0050-35.2050	Model Control (aircraft only) 1.5Watt maximum power
35.2500 -37.7500	Military (mostly army) Fixed/Mobile 50kHz channel spacing
37.5000-47.0000	Mobile (extensively military vehicles and manpacks), Radio Astronomy, ISM, cordless telephones, Television broadcasting & model control
37.7500-38.2500	Cambridge Observatory (astronomy)
37.7500 -40.0000	Military mobile 50kHz channel spacing
39.9150 -40.1200	Some beacons on space satellites have used this sub-band

From-to Pairing	Allocation
39.9375/31.0375	CT1 cordless phones portable unit TX
39.9625/31.0625	CT1 cordless phones portable unit TX
39.9875/31.0875	CT1 cordless phones portable unit TX
40.0125/31.1125	CT1 cordless phones portable unit TX
40.0375/31.1375	CT1 cordless phones portable unit TX
40.0625/31.1625	CT1 cordless phones portable unit TX
40.0875/31.1875	CT1 cordless phones portable unit TX
40.1125/31.2125	CT1 cordless phones portable unit TX
40.0500	Military Distress Frequency
40.6650-40.9550	Model control (100mW maximum)
40.6800	Industrial, Scientific & Medical
41.0000-47.4500	Military tactical mobile (vehicles/manpacks/data) 50kHz channel spacing
41.0000-68.0000	Television Broadcasting (not UK) band I
46.6100-46.9700	Unapproved cordless telephone handsets (US system B & NFM)
47.309375	Long range security alarms
47.318750	Long range security alarms
47.331250	Long range security alarms
47.356250	Long range security alarms
47.4000	Vehicle paging alarms
47.41875/77.5500	Extended range CT1 cordless phones
47.43125/77.5125	Extended range CT1 cordless phones
47.45625-47.54375	Cordless telephone handsets (8 Channels NFM, base transmits 1.6240-1.7829)
47.6800-50.0000	Land Mobile, Broadcasting, Amateur & baby listeners, walkie-talkies, wireless microphones & cordless 'phones (all NFM)
48.9750	On-site radio paging
48.9875	On-site radio paging
48.9900-49.6800	Unapproved long range cordless phone bases (22 Channels NFM, portables transmit 69.7200-70.2750
49.0000-49.8750	Private paging systems
49.0000 -50.0000	Unapproved devices mostly intended for use
	in USA
49.4250-49.4750	Hospital paging systems
49.6700-49.9700	Unapproved cordless telephone bases (US system B NFM)
49.8200-49.9000	Low powered radio control toys, baby alarms, walkie talkies etc
49.8300-49.8900	Unapproved cordless telephones
50.0000-52.0000	6 Metre Amateur band (UK allocation & NFM, CW & SSB)
50.0000-54.0000	6 Metre Amateur band (US allocation & NFM, CW & SSB)
50.5000	Video transmissions railways track to train using leaky feeder
52.0000-60.0000	Land Mobile and radio microphones.
53.8000-55.6000	BBC high powered (4 Watt) radio microphones
60.0000-64.0000	Radio Microphones
60.8000-62.6000	BBC radio microphones. 100 kHz channel spacing
64.0000-68.0000	Fixed & Land Mobile including Military
68.0000-70.0250	Land Mobile & Repeaters (military)
69.3000	Spot Frequency for Sea Cadets (AM)
69.8250-69.9750	Outside Broadcast camera links (talkback)
70.0000-70.0500	Unapproved cordless 'phones
70.0250-70.5000	4 Metre Amateur Band. CW, NFM and SSB in use.
70.5000-71.5000	Land Mobile (emergency services)
70.5125-71.5000	Fire service bases (mobiles transmit 80.0000-84.0000)
71.5000-72.8000	Low Band PMR mobiles (bases transmit +13.5MHz)
72.3750 85.8750	Short-term hire mobile
72.5250-72.7000	Ambulance bases in some areas (mobiles transmit 86.0250-86.2000)
72.5375/86.0375	Private Ambulances National Network (Mobiles)
72.8000-74.8000	Land mobile (Government)
72.8000-73.7000	Military simplex channels using 25kHz spacing

From-to Pairing Allocation

73.7000 -74.7875	Military (RAF ground services)
74.8000-75.2000	Aeronavigation guard band
75.0000	Approach fan beams, inner, middle & outer markers (AM)
75.2000-76.7000	Outside broadcast links and military, mostly allocated to USAF British bases (NFM)
75.2000-75.3000	BBC outside broadcast links
76.7000-78.0000	Fixed and Land mobile (PMR and government)
76.9625-77.5000	Fixed and Mobile. Government, PMR mobiles (paired with 86.9625-87.5000)
77.5125/47.43125	Extended range CT1 cordless phones
77.5500/47.41875	Extended range CT1 cordless phones
78.0000-80.0000	Land Mobile. Government and private users.
78.1000	Air Training Corps (nation-wide)
78.1875	BBC OB and engineering
78.2000	BBC OB and engineering
78.2125	Microwave link setting-up channel (nation-wide) and BBC OB crews
78.2225	BBC OB and engineering
78.2375	BBC OB and engineering
78.2500	BBC OB and engineering
79.0000-80.0000	RAF ground services/ police
80.0000-84.0000	Land Mobile & Fixed (extensive emergency service use).
80.5000-82.5000	Radio Astronomy (Cambridge University)
80.0000-84.0000	Fire Mobiles (paired with 70.5000-71.5000)
80.0125	Fire tender intercommunication (Channel 21)
80.0750	Fire tender intercommunication (Channel 22)
81.9500	Fire operations London
83.9960-84.0040	ISM
84.0000-85.0000	Fixed & Land Mobile (mostly military)
84.1250-84.3500	RAF ground service bases (paired with 73.7000-73.9250)
84.3000	RAF Mountain rescue teams (single Frequency simplex)
84.0000-85.0000	Military
85.0000-87.5000	Low Band PMR Bases.

From-to Pairing Allocation

85.0125-86.2875	PMR, Community Repeaters (paired with 71.5125-72.7825)
85.1375-85.2000	British Telecom engineering channels
85.8500/72.2500	National radio engineering channel
85.8750/72.3750	Low Band demonstration and short term hire channel bases
86.0375/72.5375	Private Ambulance bases National Network
86.137/ 72.6375	National engineering channel
86.3000-86.7000	Single Frequency Simplex channels
86.3125	National Mountain Rescue channel 1
86.3250	Red Cross Channel 1
86.3500	National Mountain Rescue channel 2 (in some areas, Red Crossand lifeguards)
86.3625	Scouts national channel 1
86.3750	REACT emergency teams (nationwide)
86.4125	Red Cross/mountain rescue channel 2
86.4250	Forestry Commission channel 3
86.4375	Motor Rally Safety channel
86.4500	Forestry Commission channel 2
86.4625	County Councils
86.4750	British Rail National Incident Channel
86.5000	Nuclear Spills Teams channel 1
86.5250	Nuclear Spills Teams channel 2
86.5500	Nuclear Spills Teams channel 3
86.5750	NCB mine rescue teams
86.6250	Scouts national channel 2
86.6750	Nuclear fire and radiation check teams
86.7000	BNF nuclear hazard check teams
86.9625-87.5000	Split Frequency Simplex bases (paired with 76.9625 77.5000)
87.5000-108.0000	FM Broadcast Band (Band II).
108.0000-117.9750	Aeronautical Radionavigation beacons including VHF Omnirange (VOR) and Doppler VOR (DVOR). Beacons identified by a three letter code in CW. Some of those located at or near airfields carry AM voice information on weather/runway/

From-to Pairing Allocation

From-to Pairing	Allocation
	warnings etc. This service is known as Aerodrome Terminal Information Service (ATIS).
118.0500-136.9750	International Aeronautical Mobile Band. This is the VHF band used by all civilian and some military airfields. It is subdivided into 760 channels with 25 kHz spacing but 8.33kHz channel spacing is being increasingly used, mode is AM.
118.0000-123.0000	Mostly control tower frequencies
121.5000	International Distress Frequency
123.0000-130.0000	Mostly airways frequencies (some ground & approach control)
123.1000	Search & Rescue (SAR)
130.0000-132.0000	Mostly company frequencies (airline crews to ground staff)
132.0000-136.0000	Mostly airways
135.5500-135.6450	Sub-band was once used for the American ATS series satellites
137.0000-138.0000	Space to Earth Communications & Weather Imaging Satellites (see satellite sub-section for full details)
138.0750-138.1750	Paging Systems.
138.0000-138.2000	USAF bases in some areas
140.96875	Short-term hire channel (single Frequency simplex)
141.0000-141.9000	Land Mobile mostly used by BBC, Independent Television and Radio for outside broadcast links, radio cars, etc. All Single Frequency Simplex using NFM
141.0000-141.2000	Mostly ITV
141.2000-141.9000	Mostly BBC
141.9000-143.0000	Mobile (Government) including, land, air & space satellite communications
142.0000-143.0000	Air-to-Air and Air-to-Ground. Sub-band used fairly extensively by military in continental Europe but rarely in UK

From-to Pairing Allocation

From-to Pairing	Allocation
142.7200	USAF Air-to-Air
142.8200	USAF Air-to-Air
144.0000-146.0000	Amateur 2 Metre Band including satellite allocation. CW, SSB & NFM used
146.0000-148.0000	Land Mobile & Fixed.
147.8000	Used in many areas for Fire Brigade alert pagers
148.0000-149.0000	Fixed and Land Mobile
148.56000	NOAA-series satellite telecommand uplink
149.0000-149.9000	Mobile (Military). Used particularly by USAF and RAF
149.8500	Common channel at many military bases
149.9000	Air Training Corps nationwide (channel 2)
149.9000-150.0500	Radionavigation by satellite. Doppler shift position fixing, paired with 399.9-400.05
150.0500-152.0000	Radio Astronomy and Oil Slick Markers
150.1100-150.1850	Slick Markers
152.0000-153.0000	Land Mobile
153.0000-153.5000	National and local area radio paging systems
153.5000-154.0000	Land Mobile (military) and meterological aids
156.0000-174.0000	Fixed & Mobile (land and Marine). The Marine VHF service falls within this band which also includes message handling services and mobile telephone systems. All NFM
156.0000	Marine channel '0'. Lifeboats and Coastguard
156.0000-157.4250	Marine channels single and split Frequency Simplex
156.8000	Marine Distress & Calling channel 16
157.4500-158.4000	Private Marine channels & message handling services
158.4000-158.5250	Private and Dockside using Simplex
158.5375-159.9125	Digital Pacnets
159.2500-160.5500	Private channels and message handling services, some PMR channels
160.5500-161.0000	Marine Channels
161.0000-161.1000	Paging systems acknowledge (Paired with 459.1000-459.5000)
161.1250-161.5000	Private Marine channels
161.5000-162.0500	Marine channels

From-to Pairing Allocation

From-to Pairing	Allocation
162.0500-163.0000	Private channels and message handling services
163.0375-164.4125	Digital Pacnets
163.9000/159.4000	PMR (not always split)
163.9250/159.4250	PMR (not always split)
163.9875	PMR
164.0000	PMR
164.0875	PMR
164.1250	PMR
164.1875	PMR
164.4375-165.0375	Private message handling services including paging and telephone patching
165.0625-168.2500	High Band PMR bases.
169.8625-173.0500	High Band PMR bases
166.1000-166.6125	Extensively used by ambulances
167.2000/172.2000	High Band demonstration & short term hire channel bases
167.9920 -168.0080	Industrial, scientific & Medical
168.24375-168.30525	PMR simplex channels
168.2875	Local authority alarms spot Frequency
168.84375-169.39375	High Band PMR Simplex channels
168.9375	Local authority alarms spot Frequency
168.9750	BBC Engineering
169.0125-169.7625	Short term hire channels (single Frequency simplex) all NFM
169.4125-169.8125	Pan-European paging system (ERMES)
169.81875-169.84375	PMR Simplex channels.
172.000/167.2000	Short term hire mobiles, PMR demo channel.
170.4500-170.8000	Private Security Firms mobiles (paired with 166.0000-165.8500)
170.9000-171.4250	Ambulance mobiles
173.04375-173.09375	PMR simplex channels
173.1000-173.2000	Low powered devices
173.2000-173.3500	Low Powered telemetry and telecontrol
173.3500-173.8000	Radio deaf aids, Medical and biological telemetry
173.8000-175.0000	Radiomicrophones
174.5000-225.0000	Land Mobile, Fixed, Radiolocation & Radiomicrophones, Digital Audio Broadcasting, and Television Broadcasting (not UK)
225.0000-328.6000	Aeronautical mobile (military) using AM simplex, ground-to-air, air-to-air, tactical, etc. Some satellite allocations

From-to Pairing Allocation

From-to Pairing	Allocation
235.0000-273.0000	Extensively used for military satellite downlinks (FleetSatcom West etc)
243.0000	Military distress Frequency. Life-raft beacons, SARBE's, PIRBs, etc. Frequency monitored by COSPAS/ SARSAT satellites
257.8000	Common airfield Frequency
259.7000	NASA Shuttles (AM voice)
296.8000	NASA Shuttles (AM voice particularly used on 'spacewalks')
326.5000-328.5000	Radio astronomy (Jodrell Bank)
344.0000	Common airfield Frequency
362.3000	Common airfield Frequency
328.6000-335.4000	Aeronautical radionavigation - ILS glideslope beams paired with VORs in the 108-118MHz band
335.4000-399.9000	Aeronautical mobile (military) using AM simplex, ground-to-air, air-to-air, tactical, etc
399.9000-400.0500	Radionavigation by satellite, paired with 149.9-150.05MHz band
400.0000-400.1500	Standard Frequency and time signal satellites
401.0000-406.0000	Fixed and mobile, meteorological satellites, space-earth communications
401.0000-402.0000	Space-Earth coms.
401.0000-403.0000	Meteorological sondes & satellites
401.0000-405.0000	Military telemetry links
406.0000-406.1000	Mobile satellite space-earth communications
406.05000	Emergency locator beacons (identification and location by satellite)
406.1000-410.0000	Fixed and mobile (Government), radio astronomy & radio positioning aids
406.5000-409.0000	North Sea oil rig positioning aids
410.0000-415.0000	TETRA PMR, shared with government services
415.0000-420.0000	Government fixed, mobile and space research
420.0000-450.0000	Fixed, Mobile, Amateur & radiolocation

From-to Pairing Allocation

From-to Pairing	Allocation
422.0000-425.0000	Military & Radio altimeters
425.0250-425.4750	PMR mobiles (Paired with 445.5250-445.9750)
425.5250-428.9750	PMR bases (Paired with 440.0250-443.4750)
429.0000-431.0000	Military & Radiolocation
431.00625-431.99375	PMR mobiles (London only, paired with 448.00625-448.99375)
430.0000-440.0000	70cm Amateur Band, SSB, NFM, CW, RTTY, Packet, etc, & Military
440.0250-443.4750	PMR bases (Paired with 425.5250-428.9750)
443.5000-445.5000	Military & radiolocation
445.5250-445.9750	PMR bases (Paired with 425.4750-425.0250)
446.0250-446.4750	PMR simplex 12.5kHz spacing
446.00625-446.09375	PMR446 Public simplex radio, 8 channels 12.5kHz spacing with 6.25kHz offset using 500mW handportables
446.4750-452.250	Fire channel 02
448.00625-448.99375	PMR Bases (London area only, paired with 431.00625-431.99375)
449.7500-450.0000	Earth-Space Telecommand
450.0000-470.0000	Fixed & Mobile (including marine). Mostly PMR with some emergency services, Paging, telemetry, etc
451.4000	Fire brigade on-site handhelds Ch1
451.4500	Fire brigade on-site handhelds Ch 2
453.0250-453.9750	PMR Bases (Paired with 459.5250-460.4750)
454.0125-454.8375	Wide area paging systems
455.0000-455.5000	BBC, ITV, ILR Base units for OB's (some to units paired with mobiles at +5.5MHz) Airport ground services including tower relays (typically 455.4750 455.9750 etc)
455.5000-456.0000	Some PMR (Scotland) & airport ground services
456.0000-456.9750	PMR Bases (extensively used at airports, paired with 461.50000-462.4750)
456.9250/462.4250	Short-term hire bases
457.0000-457.5000	Point-to-point links (Paired with 462.5000-463.0000)
457.50625-458.49375	Scanning telemetry (Paired with 463.00625-463.99375)
457.5250/467.5250	On-board-ship communications (international)
457.5500/467.5500	On-board-ship communications (international)
457.5750/467.5750	On-board-ship communications (international)
457.5250/467.7500	On-board-ship communications (US/Canada system)
457.5500/467.7750	On-board-ship communications (US/Canada system)
457.5750/467.8000	On-board-ship communications (US/Canada system)
457.6000/467.8250	On-board-ship communications (US/Canada system)
458.5000-459.5000	Model control, paging, telemetry & local communications
458.5000-458.8000	Low power (500mW) telemetry
459.1000-459.5000	On-site paging systems (paired with VHF 161.0000-161.1000, return 'acknowledge' signal)
459.5250-460.4750	PMR Mobiles (Paired with 453.0250-453.9750)
460.5000-461.5000	Point-to-point links, airport ground services and broadcast engineering (Paired with 467.0000-468.0000)
461.5000-462.4750	PMR Mobiles (Paired with 456.0000-456.9750)
462.4250/456.9250	Short-term hire channel mobiles
462.4750	Long-term hire (single Frequency simplex)
462.5000-463.0000	Point-to-Point links (Paired with 457.5000-458.5000)
463.0000-464.0000	Telemetry links (Paired with 457.5000-458.5000
466.0625-466.0875	Wide area paging
467.0000-467.8250	Simplex point-to-point & ILR broadcast links, also on-board-ship comms
467.5250 457.5250	On-board-ship comms.(international)
467.5500 457.5500	On-board-ship communications (international)
467.5750 457.5750	On-board-ship communications (international)

From-to Pairing Allocation	
467.7500 457.5250	On-board-ship communications (US/Canada system)
467.7750 457.5500	On-board-ship communications (US/Canada system)
467.8000 457.5750	On-board-ship communications (US/Canada system)
467.8250 457.6000	On-board-ship communications (US/Canada system)
467.8250-468.0000	Point-to-point links
468.5000-469.0000	PMR, outside broadcast links, model control.
469.0000-470.0000	Outside Broadcast link talkback and mobiles
470.0000-854.0000	U.K. Band IV Television broadcasting, Studio talkback systems, Radio Astronomy & Aeronautical radionavigation
471.0000-585.0000	Television broadcasting Band IV
537.0000-544.0000	Studio talkback mobiles
716.0000-725.0000	Studio talkback mobiles
590.0000-598.0000	Aeronavigation ground radar
614.0000	Radio Astronomy (Cambridge & Jodrell Bank)
610.0000-890.0000	Television Broadcasting Band V
702.0000-726.0000	Soviet direct TV broadcast satellites
716.0000 725.0000	Studio talkback bases
800.000-1000.000	Molniya communications satellites (data and NFM)
862.000-870.000	Fixed and mobile (not aeronautical)

From-to Pairing Allocation	
863.000-865.000	Cordless headphones
870.0000-889.0000	Fixed & Mobile(mostly military), Industrial, scientific & medical & Anti-theft devices.
886.0000-890.0000	Industrial, scientific & medical
888.0000-889.0000	Anti-theft devices (500mW maximum)
915.0000-935.0000	Fixed & Mobile (Government) & Space communications
960.0000-1215.0000	Aeronavigation (Distance measuring equipment - DME) & TACANS (radar transponders - IFF)
1215.0000-1240.0000	Radiolocation and radionavigation by satellite
1240.0000-1296.0000	Radiolocation
1296.0000-1300.0000	Amateur Radio 23cm band (NFM, SSB, WBTV, etc)
1300.0000-1365.0000	Amateur Radio 23cm band & Radiolocation (government)
1365.0000-1427.0000	Radiolocation, Space research & satellite exploration
1400.0000-1427.0000	Earth exploration satellites, astronomy & space research
1427.0000-1429.0000	Fixed & Mobile (Government) & Earth-Space satellite links
1429.0000-1450.0000	Fixed & Mobile (Government)
1450.0000-1525.0000	Fixed & Mobile (telephony, telecontrol & telemetry)

Glossary of abbreviations and definitions

Aeronautical Distress frequencies; allocated solely for use by aircraft in distress.

Aeronautical mobile; Allocations for communication between aircraft and ground stations. The main international band lies between 118-137MHz.

Aeronautical Radionavigation; Radio beacons for aircraft navigation. They include VHF omni-range (VOR), doppler VOR (DVOR), distance measuring equipment (DME), instrument landing systems (ILS), tactical navigation (TACAN), outer, middle and inner fan markers (OM, MM, IM), etc.

Aeronautical search and rescue frequencies allocated solely for aircraft involved in search and rescue (SAR) duties.

Amateur; the amateur service is for use by licensed individuals for the purpose of self-training and experimentation.

Astronomy; frequencies allocated for research into radio emissions from sources such as other galaxies.

Broadcast; Transmissions intended for reception by a large group or even the general public.

BT; British Telecom.

Citizens Band; A low powered communications service available to the public.

Cordless 'phone; A telephone handset that does not require direct connection to the exchange line.

COSPAS/SARSAT; Joint US, USSR, Canadian and French rescue service using weather satellite to fix the position of emergency rescue beacons.

ELINT; Electronic intelligence gathering (typically spy satellites).

Emergency service; Allocations for police, fire and ambulance services.

EPIRB; Emergency position indicating rescue beacon.

Fixed; A base station linked to another base station or non-mobile facility such as a repeater. Often known as point-to-point services.

FSK; Frequency shift keying.

IFF; Identify & friend or foe.

ILR; Independent local radio.

ISM; Industrial, scientific and medical. These allocations are for equipment which use radio waves to function. These allocations are not for communication purposes.

Land Mobile; Communications between a fixed base and mobile or portable equipment or between the mobile stations themselves.

Locator; The transmission of signals for navigation, position fixing and tracking.

Maritime Mobile; Services for ship-to-shore and ship-to-ship communications.

Message handling; Similar to PMR but many stations operating through a central operator at a base station.

Meteorology; The transmission of weather data from remote platforms such as sondes, buoys or satellites to ground stations.

Military; British military allocations cover the army, Royal Air Force, Royal Navy, military police and United States Air Force (USAF).

Mobile; Any mobile service. Air, marine or land.

Mobile satellite service; Communication between a mobile station and satellite (usually the satellite is acting as a relay or repeater to a distant ground station).

MOD; Ministry of Defence.

Model Control; The use of radio signals to control the movement of model boats, aircraft and cars.

NOAA; National Oceanic and Atmospheric Administration (USA).

On-site paging; A paging service operating in a restricted area such as a hospital, factory or hotel.

Pager; A miniature radio receiver which emits a tone when it receives a signal with its individually assigned code.

Positioning aid; A beacon used to emit a transmission for precise positioning or navigation. Often used for positioning such things as oil rigs.

PMR; Private mobile radio. Allocations for non-government users for communication between base stations and mobile units.

PMR446; Pan-European public radio system allocation in the 446MHz range using 500mW handhelds with 500mW maximum radiated power

Radio altimeter; The use of radio signals to measure the height of an aircraft above ground.

Radio microphone; A microphone used in broadcast studios, theatres and the film industry where the unit transmits the sound as a low powered radio signal which is picked-up by a remote receiver and then amplified.

SARBE; Search and rescue beacon. A small radio beacon attached to a lifejacket or dinghy.

Satellite navigation; Position fixing by reference to transmissions from a satellite.

Selcal; Selective calling system where a receiver only activates when it receives a pre-determined code.

Slick marker; A low powered floating beacon used to check the movement of oil slicks.

Standard Frequency; Transmission from a highly stable transmitter which is accurate enough to be used for calibration and reference. The signals often include coded signals of highly accurate time as well.

Telecontrol; A signal containing command information to control remote equipment.

Telemetry; A radio signal containing data in coded form.

Television; A radio signal containing visual images.

TETRA; TErrestrial Tunked Radio, a digital radio system used for emerency services across Europe and for civil trunked PMR

Weather satellite; A space satellite that sends weather pictures back to an earth station.

Wide area paging; A paging service not confined to a private site.

Aeronautical bands

Aeronautical and marine bands, unlike all other bands, are standard world wide. Aircraft transmissions are of two types: civilian and military. Civilian aircraft transmissions use two bands: HF using SSB for long distance communication, and VHF for communications up to distances of several hundred miles, Military communication primarily uses UHF, and all VHF/UHF communications (civilian and military) are AM (amplitude modulated). A list of civilian and military airports and bases and their and corresponding transmission frequencies are shown, but it should be noted that they may be subject to change

---

British and Irish Civil (civ) airports and Military (mil) air/ground stations.

Airfield

Aberdeen (Dyce)
EGPD civ 7 miles NW of Aberdeen
MET 125.725 (Scottish Volmet)
118.300 (Kirkwall Met)
ATIS 121.850, 114.300
APPROACH 120.400
TOWER 118.100
GROUND 121.700
VDF 120.400, 121.250, 128.300
RADAR 120.400, 128.300
FIRE VEHICLES 121.600

Aberporth
EGUC mil 5 miles NE of Cardigan
AFIS 122.150, 259.000

Abingdon
EGUD mil 5 miles SW of Oxford
TOWER 130.250, 256.500
SRE 122.100, 123.300, 120.900, 256.500

Alconbury
EGWZ mil 4 miles NW of Huntingdon
ATIS 231.175
MATZ see Wyton
TOWER 122.100, 383.45, 257.800, 315.100
GROUND 259.825
DISPATCH 342.225
DEPARTURE 134.050, 375.535
COMMAND POST 278.050, 340.125
METRO 358.600, 284.925

Alderney
EGJA civ Channel Islands
TOWER 125.350
APPROACH 128.650 (Guernsey)

Andrewsfield
EGSL civ near Braintree (Essex)
A/G 130.550

Audley End
EG civ near Saffron Walden (Essex)
A/G 122.350

Badminton
EG civ 6 miles NE of Chipping Sodbury
A/G 123.175
Bagby civ Thirsk
A/G 123.250

Airfield

Baldonnel/Casement
EIME civ Republic of Ireland
APPROACH 122.000
TOWER 123.500
GROUND 123.100
RADAR 122.800 122.300 (Dublin Military)
PAR 129.700

Bantry
EI civ Republic of Ireland
A/G 122.400

Barra
EGPR civ Traigh Mhor (Western Isles)
AFIS 130.650

Barrow
EGNL civ North end of Walney Island
A/G 123.200
TOWER 123.200

Barton
EG 5 miles W of Manchester
A/G 122.700

Battersea Heliport
EGLW civ River Thames at Battersea
TOWER 122.900

Beccles Heliport
EGSM civ 2 miles SE of Beccles (Suffolk)
A/G 134.600

Bedford
EGVW mil 5 miles N of city at Thurleigh
MATZ 124.400
APPROACH 130.700, 124.400, 265.300, 277.250
TOWER 130.000, 337.925
VDF 130.700, 130.000, 124.400, 277.250
PAR 118.375, 356.700

Airfield

Belfast Aldergrove
EGAA civ 13 miles NW of Belfast
APPROACH 120.000, 310.000
TOWER 118.300, 310.000
DISPATCHER 241.825
GROUND 121.750
VDF 120.900
RADAR 120.000, 120.900, 310.000
FIRE VEHICLES 121.600

Belfast City
EGAC civ 2 miles E of city centre
APPROACH 130.850
TOWER 130.750
SRE 134.800

Bembridge
EGHJ civ Isle of Wight
A/G 123.250

Benbecula
EGPL civ
APPROACH 119.200
TOWER 119.200

Benson
EGUB mil 10 miles SE Oxford city
MATZ 120.900
APPROACH 120.900, 122.1, 362.3, 358.800
TOWER 122.100, 279.350
GROUND 340.325
VDF 119.000
SRE 119.000

Bentwaters
EGVJ mil 6 miles NE Woodbridge (Suffolk)
ATIS 341.650
MATZ 119.000
APPROACH 119.000, 362.075
TOWER 122.100, 264.925, 257.800
GROUND 244.775
DISPATCH 356.825
DEPARTURE 258.975
SRE/PAR 119.000, 362.075
COMMAND POST 386.900 **due for closure**

Biggin Hill
EGKB civ 4 miles N Westerham (Kent)
ATIS 121.875
APPROACH 129.400
TOWER 138.400
RADAR 132.700 (Thames)

Birmingham
EGBB civ 6 miles SE of city at Elmdon
ATIS 120.725
ATC 131.325
APPROACH 131.325
TOWER 118.300
GROUND 121.800
VDF 131.325
RADAR 131.325, 118.050
FIRE VEHICLES 121.600

Airfield

Blackbushe
EGLK civ 4 miles W Camberley (Hants)
AFIS 122.300

Blackpool
EGNH civ South of Town at Squire's Gate
APPROACH 135.950
TOWER 118.400
VDF 135.950, 118.400
SRE 119.950

Bodmin
EG civ 2 miles NE Bodmin (Cornwall)
A/G 122.700

Booker see Wycombe

Boscombe Down
EGDM mil 5 miles N Salisbury
MATZ 126.700, 380.025
ATIS 263.500
APPROACH 126.700, 276.850, 291.650
TOWER 130.000, 370.100
PAR 130.750
SRE 126.700

Boulmer
EGOM SAR 8 miles E of Alnwick
A/G Boulmer Rescue 123.100, 254.425, 282.800, 299.100

Bourn
EGSN civ 7 miles W Cambridge
A/G 129.800

Bournemouth
EGHH civ 4 miles NE Bournemouth
ATIS 121.950
APPROACH 119.625
TOWER 125.600
GROUND 121.700
RADAR 119.625, 118.650
FIRE VEHICLES 121.600

Bridlington
EG civ
A/G 123.250

Bristol
EGGD civ 7 miles SW Bristol
ATIS 121.750
APPROACH 132.400
TOWER 133.850
VDF 132.400
SRE 124.350

Brize Norton
EGVN mil 5 miles SW Witney (Oxon)
MATZ 119.000
ATIS 235.150
APPROACH 133.750, 119.000, 342.450, 362.300
TOWER 126.500, 257.800, 381.200
GROUND 126.500, 370.300
DIRECTOR 130.075, 382.550
RADAR 134.300, 257.100

Airfield

Brough
EG civ 6 miles W of Hull
APPROACH 118.225
TOWER 130.550

Caernarfon
EG civ
A/G 122.250

Cambridge
EGSC civ 2 miles E of city
AP/DF 123.600
TOWER 122.200, 372.450
SRE 130.750
FIRE 121.600 fire vehicles

Cardiff Rhoose
EGFF civ 12 miles SW of Cardiff
MET 128.600 London Volmet South
ATIS 119.475
AP/DF 125.850, 277.225
TOWER 125.000
RAD/PAR 125.850, 120.050

Carlisle Crosby
EGNC civ 5 miles NE of Carlisle
APPROACH/ TOWER 123.600
DF 123.600

Carrickfin
EI civ Republic of Ireland
A/G 129.800

Chichester
EGHR civ N of Chichester (Sussex)
APPROACH 122.500
TOWER 120.650
A/G 122.450
VDF 122.450

Chivenor
EGDC mil 4 miles W of Barnstaple
(Devon)
MATZ/APP 130.200, 122.100,
362.300, 364.775
TOWER/GROUND 122.100,
362.450
GROUND 122.100, 379.925
VDF 130.200
PAR 123.300
SRE 122.100, 362.300

Church Fenton
EGXG mil 6 miles NW of Selby
(Yorks)
MATZ/APPROACH 126.500,
282.075, 362.300
TOWER 122.100, 262.700, 257.800
GROUND 122.100, 340.200
PAR 123.300
SRE 231.00, 362.300
DEPARTURE (Linton) 129.150,
381.075, 292.800

Clacton
EG civ West of Clacton (Essex)
A/G 122.325

Colerne
EG mil near Bath
A/G 122.100

Airfield

Coltishall
EGYC mil 9 miles N of Norwich
(Norfolk)
MATZ/APPROACH 125.900,
122.100, 379.275, 293.425, 342.250
TOWER 122.100, 142.290, 288.850
GROUND 269.450
VDF 125.900, 122.100, 293.425,
342.250
SRE 125.900, 123.300, 293.425,
342.250
PAR 123.300
DIRECTOR 244.750

Compton Abbas
EGHA civ 2 miles E of Shaftesbury
(Dorset)
A/G 122.700

Coningsby
EGXC mil 15 miles NE of Sleaford
(Lincs)
MATZ 120.800
APPROACH 120.800, 122.100,
312.225, 362.300
TOWER 121.100, 120.800, 275.875
GROUND 122.100, 318.150
SRE 120.800
PAR 123.300, 312.225, 362.300
DEPARTURE 344.625

Connaught
EIKN civ Republic of Ireland
TOWER 130.700
GROUND 121.900

Cork
EICK civ Republic of Ireland
MET 127.000 Dublin Volmet
APPROACH 119.900
TOWER 119.300, 121.700
GROUND 121.800

Cosford
EGWC civ 9 miles NW of
Wolverhampton
APPROACH 276.125, 362.300
TOWER 122.100, 357.125

Cottesmore
EGXJ mil 4 miles NE of Oakham
(Leics)
MATZ 123.300
APPROACH/DF 123.300, 380.950
TOWER 122.100, 130.200, 370.050,
257.800
GROUND 122.200, 336.375
PAR 123.300
SRE 123.300, 380.950
DEPARTURE 130.200, 376.575

Coventry
EGBE civ 3 miles S of Coventry
APROACH 119.250
TOWER 119.250, 124.800
GROUND 121.700
VDF 119.250, 122.000
SRE 119.250, 122.000
FIRE 121.600 fire vehicles

83

Airfield	Airfield

Cranfield
EGTC civ 4 miles E of M1 junctions
13/14
ATIS 121.875
APPROACH 122.850, 362.150
TOWER 123.200, 122.850, 341.800
VDF 122.850, 123.200, 124.550
RAD 122.850, 372.100

Cranwell
EGYD mil 4 miles NW of Sleaford
(Lincs)
MATZ 119.000
APPROACH 122.100, 119.000,
340.475, 362.300
TOWER 122.100, 379.525, 257.800
GROUND 297.900
SRE 123.300
PAR 123.300

Crossland Moor
EG civ 3 miles SW Huddersfield
A/G 122.200

Croughton
EG mil
A/G 343.600, 344.850

Crowfield
EG civ
A/G 122.775

Culdrose
EGDR mil 1 mile SE of Helston
(Cornwall)
ATIS 305.600
APPROACH/MATZ 134.050, 241.950
TOWER 122.100, 123.300, 380.225
GROUND 310.200
RADAR 122.100, 134.050, 241.950,
339.950
PAR 122.100, 123.300, 259.750,
339.950

Denham
EGLD civ 1 mile N of M1 Jct. 1
A/G 130.725

Dishforth
EGXD mil 4 miles E of Ripon
(Yorks)
MATZ Leeming
APPROACH 122.100, 379.675,
362.300
TOWER 122.100, 259.825

Dounreay Thurso
EGPY civ 8 miles W of Thurso
AFIS 122.400 (only by prior
arrangement)

Dublin
EIDW civ
MET 122.700
ATIS 118.250
APPROACH 121.100
TOWER 118.600
GROUND 121.800
SRE 119.550, 118.500, 118.600,
121.100

Dundee
EGPN civ 2 Miles W of Dundee
APPROACH/TOWER 122.900

Dunkeswell
EG civ 5 miles NW of Honiton
A/G 123.475

Dunsfold
EGTD civ 9 miles S of Guildford
(Surrey)
APPROACH 122.550, 312.625,
367.375
TOWER 124.325, 375.400
VDF 122.550, 124.325
RAD 119.825, 122.550, 291.900

Duxford
EG civ 9 miles S of Cambridge
AFIS 122.075

Earls Colne
EGSR civ 5 miles SE of Halstead
(Essex)
A/G 122.425

East Midlands
EGNX civ Castle Donnington, off
M1
MET 126.600 (London Volmet North)
APPROACH 119.650
TOWER 124.000
GROUND 121.900
VDF 119.650
SRE 124.000, 120.125
FIRE 121.600 (fire vehicles)

Edinburgh
EGPH civ 8 miles W of Edinburgh
MET 125.725 Scottish Volmet
ATIS 132.075
APPROACH 121.200, 130.400
(departing gliders, 257.800
TOWER 118.700, 257.800
GROUND 121.750, 257.800
VDF 121.200, 118.700
RAD 121.200, 128.975
FIRE 121.600 (fire vehicles)

Elstree
EGTR civ 12 miles NW London city
centre
A/G 122.400

Elvington
EGYK mil
See Church Fenton

Enniskillen St Angelo
EGAB civ 5 miles N Enniskillen (N.
Ireland)
A/G 123.200

Enstone
EG civ 5 miles SE of Chipping
Norton
A/G 129.875

Exeter
EGTE civ 4 miles E of Exeter
APPROACH/DF 128.150
TOWER 119.800
SRE 128.150, 119.050

Airfield

Fairford
EGVA mil N of Swindon
MATZ/APPROACH Brize Norton
TOWER 119.150, 357.575
GROUND 259.975
DISPATCHER 379.475
COMMAND POST 371.200, 307.800
METRO 358.600

Fairoaks
EGTF civ 3 miles N of Woking
AFIS and A/G 123.425

Farnborough
EGUF mil W of A325
EGLF civ
A/G 130.050
APPROACH 134.350, 336.275
TOWER 122.500, 357.400
PAR 130.050, 353.850
DISPATCHER 254.850

Fenland
EGCL civ Holbeach (Lincolnshire)
AFIS and A/G 122.925

Fife see Glenrothes

Filton (Bristol)
EGTG mil 4 miles N of Bristol
APPROACH 122.275, 127.975,
256.125
TOWER 124.950, 342.025
VDF 122.275
SRE 132.350

Finningly
EGXI mil SE of Doncaster
MATZ 120.350
APPROACH 120.350, 358.775
TOWER 122.100, 379.550
GROUND 340.175
SRE 120.350, 285.125, 315.500,
344.000
PAR 123.300, 383.500, 385.400

Flotta
EG civ Centre of Orkney Island
A/G 122.150

Galway (Carnmore)
EICM civ Republic of Ireland
A/G/TOWER 122.500

Gamston (Retford)
EGNE civ
A/G 130.475

Gatwick see London Gatwick

Glasgow
EGPF Civ 6 miles W of City Centre
MET 125.725 (Scottish Voomet)
135.375 (London Volmet Main)
ATIS 115.400
APPROACH 119.100
TOWER 118.800
GROUND 121.700
RADAR 119.100, 119.300, 121.300
FIRE VEHICLES 121.600

Glenrothes (Fife)
EGPJ civ
A/G 130.450

Goodwood see Chichester

Airfield

Gloucester (Staverton)
EGBJ civ Gloucester and
Cheltenham
APPROACH 125.650, 120.970
TOWER 125.650
VDF 125.650, 122.900
SRE 122.900
FIRE VEHICLES 121.600

Great Yarmouth
EGSD gov North Denes
A/G 120.450, 122.375
HF A/G 3.488, 5.484 MHz

Guernsey
EGJB civ 3 miles S of St Peter Port
ATIS 109.400
APPROACH 128.650
TOWER 119.950
GROUND 121.800
VDF 128.650, 124.500
SRE 118.900, 124.500

Halfpenny Green
EGBO civ 6 miles W of Dudley
AFIS 123.000
GROUND 121.950

Hatfield
EGTH civ 2 miles S Welwyn
Garden City
APPROACH 123.350, 343.700
TOWER 130.800, 359.450
SRE 123.350, 119.300, 343.700

Haverfordwest
EGFE mil 2 miles N of town
AFIS 122.200

Hawarden
EGNR civ 4 miles W of Chester
APPROACH 123.350
TOWER 124.950, 336.325
VDF 123.350, 129.850
RADAR 129.850

Heathrow see London Heathrow

Henstridge
EGHS civ Somerset S of A30
A/G 130.275

Hethel
EGSK civ 7 miles SW of Norwich
A/G 122.350

Honington
EHXH mil N of Bury St Edmunds
(Suffolk)
MATZ 129.050
APPROACH 129.050, 309.950,
344.000
TOWER 122.100, 283.275, 257.900
GROUND 241.975
DEPARTURE 123.300, 309.950
SRE 129.050, 254.875, 309.950,
338.975, 344.000
PAR 123.300, 358.750, 385.400

Hucknall
EGNA civ 1 miles SW of town
(Notts)
A/G 130.800

Airfield

Humberside
EGNJ civ 15 miles E of Scunthorpe
APPROACH 123.150
TOWER 118.550
VDF 123.150
FIRE VEHICLES 121.600

Inverness (Dalcross)
EGPE civ 8 miles NE Inverness
MET 125.725 (Scottish Volmet)
APPROACH/TOWER 122.600

Ipswich
EGSE civ 2 miles SE Ipswitch
A/G 118.325

Islay (Port Ellen)
EGPI civ South end of Island
AFIS 123.150

Jersey
EGJJ civ 3 miles W of St. Helier
MET 128.600 (London Volmet South)
ATIS 112.200
APPROACH 120.300
TOWER 199.450
GROUND 121.900
FIRE VEHICLES 121.600

Kemble
EGDK mil 4 miles SW Cirencester
MATZ 118.900
APPROACH 118.900, 123.300
TOWER 118.900

Kinloss
EGQK mil (3 miles NE of Foress
(Grampian)
MET 118.300
MATZ Lossiemouth
APPROACH 119.350, 362.300,
376.650
TOWER 122.100, 336.350, 257.800
DISPATCHER 358.475
OPERATIONS 259.825
SRE 123.300, 259.975, 311.325
PAR 123.300, 370.050, 376.525

Kirkwall
EGPA civ (Orkney
MET 118.300
APPROACH/TOWER 118.300

Lakenheath
EGUL mil (5 miles N of Barton Mills
(Suffolk)
APPROACH 123.300, 398.350
RAPCON 398.350
TOWER 122.100, 358.675, 257.800
GROUND 397.975
DISPATCHER 300.825
DEPARTURE 123.300, 315.575
COMMAND POST 269.075
METRO 257.750
SRE/PAR 123.300, 243.600,
262.925, 290.825, 338.675
SRE/PAR 149.650 (NFM)

Lands End (St. Just)
EGHC civ 6 miles W of Penzance
A/G 130.700
APPROACH/TOWER 130.700

Airfield

Lasham
EGHL civ 5 miles S of Basingstoke
A/G 122.875

Lashenden
EGHK civ 10 miles SE of Maidstone
A/G 122.000

Leavesden
EGTI civ 2 miles NW of Watford
APPROACH/TOWER 122.150
VDF 122.150
SRE 122.400

Leconfield Rescue
EGXV mil
A/G 122.100, 244.875, 282.800

Leeds-Bradford
EGNM civ Half way between Leeds
Bradford
MET 126.600 (London Volmet North)
APPROACH/VDF 123.750
TOWER 120.300
SRE 121.050
FIRE VEHICLES 121.600

Leeming
EGXE mil Northallerton (Yorks)
APPROACH 127.750, 387.800
TOWER 122.100, 382.100, 394.500
VDF 132.400, 122.100, 359.200,
362.300, 387.800
PAR 122.100, 248.000, 352.900
SRE 127.750, 339.400

Lee-on-Solent
EGUS mil 4 miles west of Gosport
TOWER 135.700, 315.650

Leicester
EGBG civ 4 miles S of Leicester
A/G/FIS 122.250

Lerwick Tingwall
EG civ Shetland Isles
A/G 122.600

Leuchars
EGQL mil 7 miles SE of Dundee
MATZ/LARS 126.500
APPROACH 126.500, 255.400,
362.300
TOWER 122.100, 258.925
GROUND 120.800, 259.850
DISPATCHER 285.025
VDF 126.500
SRE 123.300, 292.475
PAR 123.300, 268.775, 370.075

Linton-on-Ouse
EGXU mil 10 miles NW of York
MATZ/LARS 129.150, 121.100,
292.800, 344.000
APPROACH 129.150, 292.800,
362.675, 362.300
TOWER 122.100, 257.800, 300.425
GROUND 122.100, 340.025
DEPARTURE 129.150, 381.075,
292.800
SRE 129.150, 122.100, 344.000,
344.475
PAR 123.300, 129.150, 259.875,
358.525

Airfield

Liverpool
EGGP civ 6 miles SE of city
MET 126.600 (London Volmet North)
APPROACH 119.850
TOWER 118.100
RADAR 18.450, 119.850

Llanbedr
EGOD mil 3 miles S of Harlech
APPROACH 122.500, 386.675
TOWER 122.500, 380.175
RADAR/PAR/VDF 122.500,
370.300, 386.675

London City
EGLC civ London Dockland
TOWER 119.425, 118.075
RADAR 132.700 (Thames), 128.025
(City)
FIRE VEHICLES 121.600

London Gatwick
EGKK civ 28 miles S of London
MET 135.375 (London Volmet Main)
ATIS 128.475
APPROACH 125.875, 134.225
TOWER 124.225, 134.225
CLEARANCE 121.950
GROUND 121.800
RADAR 134.225, 118.600, 119.600,
129.275
FIRE VEHICLES 121.600

London Heathrow
EGLL civ 14 miles W of London
MET 135.375 (London Volmet Main)
ATIS 115.100 (Biggin , 113.750
(Bovingdon)
ATIS 133.075
APPROACH 119.200, 120.400,
119.500, 127.550
TOWER 118.700, 124.475
CLEARANCE 121.700
GROUND 121.900
RADAR 119.200, 119.500, 127.550,
120.400
FIRE VEHICLES 121.600

London Stansted
EGSS civ 30 miles N of London
MET 135.375 (London Volmet Main)
ATIS 127.175
APPROACH 125.550
TOWER 118.150
GROUND 121.700
VDF 125.550, 126.950, 118.150,
123.800
RADAR 125.550, 126.950, 123.800
FIRE VEHICLES 121.600

Londonderry
EGAE civ Northern Ireland
APPROACH 123.625
TOWER 122.850
FIRE 121.600

Airfield

Lossiemouth
EGQS mil 5 miles N of Elgin
(Grampian)
MATZ/LARS 119.350, 376.650
APPROACH 119.350, 362.300,
398.100
TOWER 118.900, 122.100, 337.750
GROUND 299.400
SRE 123.300, 259.975, 311.325
PAR 123.300, 250.050, 312.400
VDF 119.350,

Luton
EGGW civ SE of Luton
MET 128.600 (London Volmet South)
ATIS 120.575
APPROACH 129.550, 128.750,
127.300, 259.875
TOWER 119.975
GROUND 121.750
VDF 129.550, 127.300, 128.750
SRE 128.750, 127.300
FIRE VEHICLES 121.600

Lydd
EGMD civ Off B2075
APPROACH/TOWER 120.700
SRE 131.300

Lyneham
EGDL mil 8 miles SW M4 junction 16
ATIS 381.000
APPROACH 118.425, 123.400,
359.500, 362.300
TOWER 118.425, 122.100, 386.825
GROUND 118.425, 122.100,
340.175
DISPATCHER 265.950
OPERATIONS 254.650
VDF 123.400
PAR 123.300, 375.200, 385.400
SRE 123.400, 300.475, 344.000

Macrihanish
EGQJ mil 4 miles W of
Campbeltown
MATZ 125.900, 122.100, 344.525,
362.300
APPROACH 125.900, 122.100,3
44.525, 362.300
TOWER 122.100, 358.600, 257.800
PAR 123.300, 337.975, 385.400
SRE 125.900, 123.300, 259.925,
344.000

Manchester
EGCC civ 10 miles S of city centre
MET 135.375, 126.600, 127.000
(Dublin Volmet)
ATIS 128.175
APPROACH 119.400, 121.350
TOWER 118.625
GROUND 121.700, 121.850
RADAR 119.400, 121.350
FIRE VEHICLES 121.600

Manchester (Barton)
EGCB civ 6 miles W of Manchester
A/G 122.700

Airfield

Manston
EGUM mil 4 miles NW of Ramsgate
EGMH civ
MATZ 126.350
APPROACH 126.350, 122.100,
362.300, 379.025
TOWER 128.775, 122.100, 344.350,
257.800
VDF 126.350, 129.450
PAR 123.300, 118.525, 312.350,
385.400
SRE 126.350, 123.300, 338.625,
344.000

Marham
EGYM mil Near Swaffham (Norfolk)
MATZ 124.150
APPROACH 124.150, 291.950,
362.300 (Eastern Radar)
TOWER 122.100, 337.900, 257.800
DISPATCHER 241.450
OPERATIONS 312.550
VDF 124.150, 122.100
PAR 123.300, 379.650, 385.400
SRE 124.150, 293.775, 344.000

Merryfield
EG mil N of Ilminster (Somerset)
APPROACH 127.350, 276.700,
362.300 (Yeovil)
TOWER 122.100, 287.100

Middle Wallop
EGVP mil 6 miles SW of Andover
(Hants)
MATZ Boscombe Down
APPROACH 126.700, 122.100,
312.000
TOWER 122.100, 372.650
PAR 364.825
SRE 312.675

Mildenhall
EGUN mil Near Barton Mills
(Suffolk)
ATIS 277.075
APPROACH 128.900
TOWER 122.550, 258.825
GROUND 380.150
COMMAND POST 379.850, 312.450
MAINTENANCE 254.625
METRO 257.750
NAVY DUTY 142.850 (NFM)
AIR MOBILITY COM 379.850 See
also Honington for MATZ/APP/
DEPART

Mona
EG mil 10 miles W of Menai Bridge
AFIS 122.000
APPROACH 379.700
TOWER 358.750
See also Valley

Netheravon
EGDN mil Near Armesbury (Wilts)
A/G/TOWER 128.300, 253.500
APPROACH 362.225
TOWER 290.950

Airfield

Netherthorpe
EGNF civ 3 miles NW of Worksop
AFIS 123.275
A/G 123.275

Newcastle
EGNT civ 5 miles NW of Newcastle
MET 126.600 (London Volmet North)
ATIS 114.250
APPROACH 124.375, 284.600
TOWER 119.700
VDF 118.500, 119.700, 126.350
RADAR 126.350, 118.500
FIRE 121.600

Newquay See St Mawgan

Newton
EGXN mil 6 miles E of Nottingham
APPROACH 122.100, 251.725,
362.300
TOWER 122.100, 257.800, 375.425

Newtownards
EGAD civ 2 miles SE of town centre
A/G 123.500

Northampton
EGBK civ 5 miles NE of city
AFIS/A/G 122.700

North Denes See Great Yarmouth

Northolt
EGWU mil Close to Uxbridge
(London area)
ATIS 300.350
APPROACH/VDF 126.450, 344.975,
362.300
TOWER 126.450, 257.800, 312.350
OPERATIONS 244.425
SRE 375.500, 379.425
PAR 130.350, 385.400

North Weald
EGSX mil NE of Epping (Essex)
A/G 123.525

Norwich
EGSH civ 3 miles N of Norwich
MET 128.600 (London Volmet South)
APPROACH 119.350
TOWER 124.250
SRE 119.350, 118.475
FIRE 121.600

Nottingham
EGBN civ Near Tollerton
A/G 122.800

Odiham E
GVO mil 8 miles N of Alton (Hants)
ATIS 276.175
MATZ Farnborough
APPROACH 122.100, 125.250,
315.975, 362.300
TOWER 122.100, 309.625, 257.800
PAR 123.300, 385.400
SRE 386.775

Old Sarum
EG civ 2 miles N of Salisbury
(Wilts)
A/G 123.575

Airfield

Old Warden
EG civ 2 miles W of Biggleswade (Beds)
A/G 123.050

Oxford (Kidlington)
EGTK civ 6 miles NW of Oxford
ATIS 121.950
AFIS 119.800
A/G 118.875
APPROACH 125.325
TOWER 118.875
GROUND 121.750
VDF 125.325

Panshangar
EG civ 4 miles W of Hertford
AFIS 120.250

Penzance Heliport
EGHK civ 1 miles E of Penzance
A/G 118.100

Perth (Scone)
EGPT civ 3 miles NE of Perth
APPROACH/VDF 122.300
TOWER 119.800

Peterborough (Con)
EGSF civ 8 miles S of Peterborough
A/G 129.725

Peterborough (Sib)
EGSP civ
A/G 122.300

Plymouth
EGHD civ 4 miles N of Plymouth
APPROACH 133.550
TOWER 122.600
VDF 133.550, 122.600

Pocklington
EG civ
A/G 130.100
Portishead Radio EG civ
A/G 131.625

Popham
EG civ Near junction 8 of M3 (Hants)
A/G 129.800

Portland
EGDP mil 4 miles S of Weymouth
MATZ 124.150, 317.800
APPROACH 124.150, 122.100, 362.300
TOWER 122.100, 123.300, 124.150, 291.000, 362.300
SRE 124.150, 122.100, 317.800, 362.300
PAR 387.500, 362.300

Predannack
EG civ Near Lizard Point (Cornwall)
TOWER 338.975, 370.000 See also Culdrose

Airfield

Prestwick
EGPK civ 28 miles S of Glasgow
MET 125.725 (Scottish Volmet)
ATIS 127.125
APPROACH 120.550, 386.925
TOWER 118.150, 121.800
RADAR 120.550, 119.450
FIRE 121.600
ROYAL NAVY OPS 337.750

Redhill
EGKR civ 1 mile E of town (Surrey)
AFIS/TOWER 120.275

Retford (Gamston)
EGNE civ 3 miles S of Town (Notts)
A/G 130.475

Rochester
EGTO civ 2 miles S of Town (Kent)
AFIS 122.250

St Athan
EGDX mil 10 miles W of Barry (S Glamorgan)
APPROACH 122.100, 277.225, 357.175
TOWER 122.100, 257.800, 336.525
SRE 123.300, 340.100, 344.000, 380.125, 385.400

St Mawgan
EGDG mil 5 miles NE of Newquay
MATZ 126.500
APPROACH 126.500, 122.100, 125.550, 357.200, 362.300
TOWER 123.400, 122.100, 241.825
DISPATCHER 245.600
OPERATIONS 260.000
VDF 126.500, 125.550
PAR 123.300, 336.550, 385.400
SRE 125.550, 344.000, 360.550

Salisbury Plain
EG mil
A/G 122.750, 253.500

Sandown
EG civ Isle of Wight
A/G 123.500

Scampton
EGXP mil 5 miles N of Lincoln
APPROACH 312.500, 362.300
TOWER 122.100, 282.400, 257.800
GROUND 372.500
DEPARTURE 249.850, 362.300
VDF 252.525
RADAR 127.350, 357.050, 344.000

Scatsta
EGPM civ
APPROACH 123.600
TOWER 123.600
SRE 122.400
FIRE 121.600

Scilly Isles
EGHE civ St Mary's Island
APPROACH/TOWER 123.150

Seething
EG civ SE of Norwich
A/G 122.600

Airfield

Shannon
EINN civ Republic of Ireland
MET 127.000 (Dublin Volmet)
ATIS 130.950
APPROACH 121.400, 120.200
OCEANIC DEPARTURE 121.700
TOWER 118.700
GROUND 121.800
RADAR 121.400

Shawbury
EGOS mil 10 miles NE of
Shrewsbury
MATZ 124.150, 254.200
APPROACH 124.150, 276.075,
362.300
TOWER 122.100, 269.100, 257.800
GROUND 337.900
SRE 124.150, 344.000
PAR 123.300, 356.975, 385.400

Sherburn-in-Elmet
EGCL civ Sherburn (Yorkshire)
A/G 122.600

Shipdham
EG civ 4 miles S East Derham
(Norfolk)
AFIS/A/G 119.950

Sheffield
EG civ 3 miles NE of city
Shobdon EGBS civ 10 miles W of
Leominster
A/G 123.500

Shoreham
EGKA civ 1 mile W of town
A/G 123.150
ATIS 121.750
APPROACH 123.150
TOWER 125.400
VDF 123.150

Sibson see Peterborough
Silverstone
EG civ Northamptonshire
TOWER/A/G 121.075 9 (by
arrangement only)

Skegness
EG civ 2 miles N of town
A/G 130.450

Sleap
EG civ 10 miles N of Shrewsbury
A/G 122.450

Sligo
EISG civ Republic of Ireland
AFIS/TOWER/A/G 122.100

Southampton
EGHI civ 1 mile W of Eastleigh
MET 128.600 (London Volmet South)
ATIS 113.350
ATC/APPROACH 120.225, 128.850,
131.000
TOWER 118.200
RADAR 120.225, 128.850
FIRE 121.600

Airfield

Southend
EGMC civ 2 miles N of town
MET 128.600 (London Volmet South)
ATIS 121.800
APPROACH 128.950
TOWER 127.725
SRE 128.950

Stanford
EG mil 1 mile NW M20 junction 11
(Kent)
A/G OPERATIONS 307.800

Stansted see London Stansted
Stapelford
EGSG civ 5 miles N of Romford
(Essex)
A/G 122.800

Stornoway
EGPO civ 3 miles E of town
(Hebrides)
MET 125.750 (Scottish Volmet)
AFIS/APPROACH 123.500
TOWER 123.500

Strathallan
EG civ Tayside
A/G 129.900

Sturgate
EGCS civ 6 miles SE of
Gainsborough
A/G 130.300

Sumburgh
EGPB civ South of Island
(Hebrides)
MET 125.725 (Scottish Volmet)
ATIS 125.850
APPROACH/RADAR 123.150
TOWER 118.250

Swansea
EGFH civ 6 miles W of Swansea
A/G/APPROACH 119.700
TOWER 119.700

Swanton Morley
EG mil 4 miles N of Dereham
(Norfolk)
TOWER 123.500

Swinderbury
EGXS mil 8 miles NE of Newark
(Lincs)
APPROACH 283.425
TOWER 122.100, 375.300

Sywell see Northampton
Tatenhill
EGBM civ 6 miles W of Burton-on-
Trent
A/G 122.200

Teesside
EGNV civ 6 miles E of Darlington
MET 126.600 (London Volmet North)
APPROACH 118.850
TOWER 119.800
VDF/RADAR 118.850, 119.800,
128.850

Airfield

Ternhill
EG mil 4 miles SW of Market Drayton
APPROACH 124.150, 122.100, 276.825, 362.300, 365.075
TOWER 124.150, 338.825, 309.550 (Chetwynd)
RADAR 123.300, 122.100, 344.375

Thruxton
EGHO civ 6 miles W of Andover (Hants)
A/G 130.450

Tiree
EGPU civ centre of island
AFIS 122.700)

Topcliffe
EGXZ mil 4 miles SW of Thirsk (Yorkshire)
MATZ Leeming
APPROACH 125.000, 121.100, 357.375, 362.300
TOWER 125.000, 121.100, 309.725, 257.800
SRE 123.300, 344.350, 385.400

Truro
EG civ near town (Cornwall)
A/G 129.800

Unst
EGPW civ Shetland Isles
APPROACH/TOWER 130.350

Upavon
EG mil 8 miles N of Amesbury (Wilts)
TOWER 275.800

Upper Heyford
EGUA mil 6 miles W of Bicester
ATIS 242.125
APPROACH 128.550, 123.300, 364.875
TOWER 122.100, 257.800, 316.000
GROUND 375.175
DISPATCHER 277.175
DEPARTURE 364.875
COMMAND POST 357.900, 359.850
METRO 257.750, 358.600
RADAR 128.550, 122.100 **On standby for closure**

Valley
EGOV mil 6 miles SE of Hollyhead
MATZ 134.350, 268.775
SAR 282.800
APPROACH 134.350, 372.325, 362.300
TOWER 122.100, 340.175, 257.800
GROUND 122.100, 386.900
VDF 134.350
SRE 134.350, 123.200, 268.775, 282.800
DIRECTOR 337.725, 344.000
PAR 123.300, 358.675, 385.400

Airfield Airfield

Waddington
EGXW mil 6 miles S of Lincoln
MATZ 127.350, 296.750
APPROACH 312.500, 362.300
TOWER 122.100, 285.050, 257.800
GROUND 342.125
DEPARTURE 123.300, 249.850
DISPATCHER 291.150
VDF 127.350
PAR 123.300, 309.675, 385.400
SRE 127.350, 123.300, 300.575, 344.000

Warton
EGNO civ 4 miles E of Lytham/St Annes
APPROACH 124.450, 130.800, 336.475
TOWER 121.600, 130.800, 254.350
SRE 124.450, 130.800, 336.475, 254.350
RADAR 129.275, 343.700

Waterford
EIWF civ Rebublic of Ireland
A/G/TOWER 129.850
Wattisham EGUW mil 10 miles NW of Ipswitch
APPROACH 135.200,
TOWER 122.100, 343.250
SRE 123.300, 277.475
PAR 123.300, 356.175, 359.825

Wellesbourne
EGBW civ 4 miles E of Stratford-upon-Avon
A/G 130.450

Welshpool
EG civ 12 miles W of Shrewsbury
A/G 123.250

West Freugh
EGOY mil 4 miles SE of Stranraer
MATZ 130.050
APPROACH 130.050, 383.525
TOWER 122.550, 337.925
SRE 130.725, 383.525

Westland heliport
EGLW civ Battersea (London)
TOWER 122.900

West Malling
EGKM civ 4 miles W of Maidstone
A/G 130.875

Weston
EIWT civ Republic of Ireland
A/G 122.400

Weybridge
EG civ just S of town (Surrey)
A/G 122.350

White Waltham
EGLM civ 4 miles SW of Maidenhead
A/G 122.600

Wick
EGPC civ
MET 118.300 (Kirkwall)
AFIS/APPROACH 119.700
TOWER 119.700

Airfield

Wickenby
EGNW civ 10 miles NE of Lincoln
A/G 122.450
Wigtown (Baldoon)
EG civ Dumfries and Galloway
A/G 123.050
Woodbridge
EGVG mil 5 miles NE of Ipswitch
MATZ/APPROACH See Bentwaters
ATIS 336.000
TOWER 119.150, 122.100, 257.800,
291.350
GROUND 307.400
COMMAND POST 282.150
METRO 259.400 **Due for closure**
Woodford
EGCD civ 4 miles E of Wilmslow
(Lancs)
APPROACH 130.050, 126.925,
269.125, 358.575
TOWER 126.925, 130.050, 358.575,
299.975
SRE 130.750, 130.050, 269.125,
358.575
Woodvale
EGOW mil 6 miles SW of Southport
(Lancs)
APPROACH 122.100, 312.800
TOWER 119.750, 259.950
Wroughton
EGDT civ 3 miles S of Swindon
A/G 123.225

Airfield

Wycombe Air Park
EGTB civ 3 miles SW of High
Wycombe
TOWER 126.550
GROUND 121.775
Wyton
EGUY mil 4 miles NE of Huntingdon
MATZ 134.050
APPROACH 134.050, 362.375,
362.300
TOWER 122.100, 312.275, 257.800
DEPARTURE 134.050, 375.525
PAR 122.100, 292.900, 385.400
SRE 123.300, 249.550, 344.000
Yeovil
EGHG mil SW of town
APPROACH 130.180, 369.975
TOWER 125.400, 372.425
SRE 130.800
RADAR 127.350, 369.875
Yeovilton
EGDY mil 2 miles E of Ilchester
MATZ 127.350
ATIS 379.750
APPROACH 127.350, 369.875,
362.300
TOWER 122.100, 372.650
GROUND 311.325
PAR 123.300, 339.975, 344.350
SRE 123.300, 338.875, 362.300
RADAR/VDF 127.350, 369.875

What you might hear

Remember that aircraft transmissions are usually short and there may be long periods when nothing is heard on a frequency. This applies, in particular, to smaller airfields where traffic movement may be quite low. In addition to approach, control tower and radar landing instructions you may also hear a variety of other messages being passed on other frequencies in the bands. Many airlines have 'company frequencies' on which aircraft crews and ground operation staff communicate. You may also hear transmissions relating to zone, area or sector controllers. These are the people who control the movements of aircraft as they fly between airports. Different sectors have different transmission frequencies and so, to follow a particular aircraft as it moves from one sector to another, you will need to change your reception frequencies, to suit. At London Heathrow and similar large airports, the sheer volume of traffic means that instructions passed to the aircraft must be done by several controllers and so you may come across frequencies which are dealing solely with such things as instructions on taxiing on the ground.

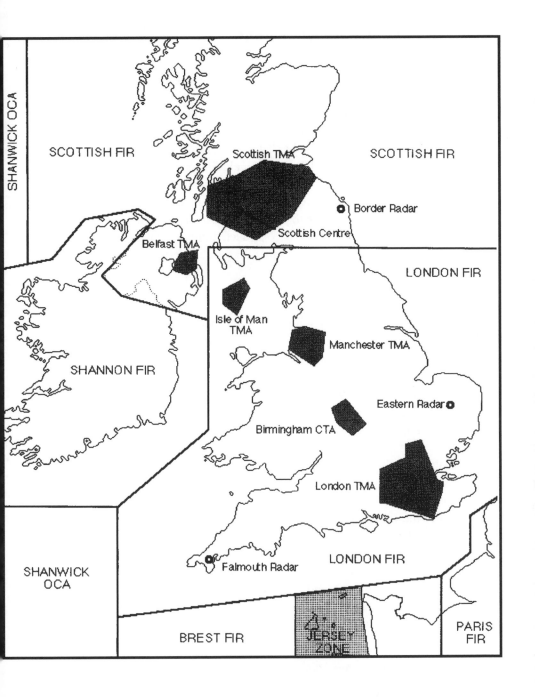

Continuous transmissions

Some frequencies are allocated solely for transmissions from the ground. The aircraft never transmit on these frequencies but the crews may listen to the broadcasts for information. The most common of these are 'VOLMETS', transmitted round the clock and detailing current weather conditions for most major airports. Automatic terminal information service (ATIS) transmissions, on the other hand, are sent out by individual airports and only include details of that airport, including current weather, runway and approach patterns in use, and any other essential information. They, in fact, contain all the information a pilot needs except actual landing permission. Pilots will listen to these transmissions and when contacting the controller will often be heard to say such things as 'information Bravo received'. The word 'Bravo' standing for the code letter which identifies the start of an ATIS transmission.

Range

Using a reasonable outside antenna it may be possible to hear ground stations up to 20 miles or so away. However, if hills or large buildings are between the scanner and the airport then this range will be considerably reduced. For instance, in my own case I cannot pick up my local airport which is only four miles away and yet can pick up another airport which is some 25 miles away in a different direction. Air-to-ground range though is a different matter altogether. Aircraft flying at tens of thousands of feet may be heard several hundred miles away even though the scanner is only operating on a small telescopic antenna. This is because the line-of-site range is greatly extended by the height of the aircraft which is transmitting from a point where there are no obstructions to block or weaken the signal. If an antenna is used solely for airband reception then it should be vertically polarised. It is worth noting, by the way, that a simple ground plane antenna of the type described in Chapter 4 is more than adequate for aircraft band-only operation. You will hear many unfamiliar expressions and considerable use of abbreviations in the airband. If you are not familiar with these, you can look them up in the airband section in Chapter 7.

United Kingdom Airways Allocations

A/way	Sector	Control
A1	Turnberry & 54.30N	Scottish Ctl 126.250 & 128.500
	54.30N & abm Stafford	London Ctl 131.050, 129.100 & 134.425
	54.30N & abm Stafford	Manchester Ctl 126.650 & 124.210
	Abm Stafford & Birmingham	London Ctl 133.700, 134.425
	Abm Stafford & Birmingham	Manchester Ctl 124.200
	Birmingham & Abm Woodley	London Ctl 133.700 & 133.975

A/way	Sector	Control
	Daventry area and	
	Birmingham Zone Ctl Birmingham	SRA/SRZ 120.500
A2	TALLA & 54.30N	Scottish Ctl 128.500
	54.30N & Abm Lichfield	London Ctl 131.050 & 134.425
	54.30N & Abm Lichfield	Manchester Ctl 126.65 & 124.200
	Abm Lichfield & Abm Birmingham	London Ctl 121.025, 133.700 & 134.425
	Abm Lichfield & Abm Birmingham	Manchester Ctl 126.65 & 124.200
	Abm Birmingham &	London Ctl 121.025, 133.700 Brookmans Park & 133.975
	South of Brookmans Park	London Ctl 127.100 & 132.450
A20	FIR Boundary & Biggin	London Ctl 127.100
	Biggin & Abm Birmingham	London Ctl 121.025, 133.700 & 133.975
	Abm Birmingham & Pole Hill	London Ctl 131.050 above FL155
	Abm Birmingham & Pole Hill	Manchester Ctl 124.200 & 126.650
A25	Dean Cross & 54.30N	Scottish Ctl 126.250 & 128.500
	54.30N & REXAM	London Ctl 128.050, 129.100 & 134.425
	54.30N & REXAM	Manchester Ctl 133.050 & 125.100
	REXAM & Cardiff	London Ctl 131.200
	Cardiff & 50.00N	London Ctl 132.600 & 135.250
	50.00N & Channel Isles Boundary	Jersey Zone 125.200
A30	London FIR	London Ctl 127.100
A34	London FIR	London Ctl 127.700 & 124.275
A37	Entire route	London Ctl 129.600, 127.950, 133.450 & 133.525
A47	Pole Hill & Lichfied	London Ctl 131.050 above FLI55
	Pole Hill & Lichfield	Manchester Ctl 126.65, 124.200 below FL175
	Lichfield & abm Birmingham	London Ctl 133.700 Above FL135
	Lichfield & abm Birmingham	Manchester Ctl 124.200 below FL175
	Abm Birmingham & Woodley	London Ctl 133.700, 121.020
	Daventry CTA below FL130	London Ctl 133.975
	South of Woodley to FIR Boundary	London Ctl 127.700, 135.050 & 124.275
B1	West of Wallasey	London Ctl 128.050, 129.100 & 134.425
	West of Wallasey	Manchester Ctl 133.050 below FL175
	Wallasey & BARTN	London Ctl 128.050 & 134.425
	Wallasey & BARTN	Manchester Ctl 125.100 below FL175
	BARTN & Ottringham	London Ctl 131.050 & 134.425
	BARTN & Ottringham	Manchester Ctl 126.650 & 124.200
	East of Ottringham	London Ctl 134.250, 127.950 & 133.525
B2	North of TMA	Scottish Ctl 124.500
	South of TMA	Scottish Ctl 135.675
B3	Belfast & 5W	Scottish Ctl 135.675
	5W & Wallasey	London Ctl 128.050 & 129.100
	Wallasey & Stafford	London Ctl 128.050 & 129.100
	Wallasey & Stafford	Manchester Ctl 125.100 & 124.200
	Stafford & abm Birmingham	London Ctl 133.7 & 121.025 above FL135
	Stafford & abm Birmingham	Manchester Ctl 125.1 & 124.200 below FL175
	Daventry CTA within A1	Birmingham Zone Ctl on 120.500 below FL80
	Abm Birmingham & Brookmans Park	London Ctl 133.700, 121.025 & 133.975
	South of Brookmans Park to London Ctl	127.100 FIR boundary 134.900
B4	Detling & Brookmans Park	London Ctl 127.100 & 134.900
	Brookmans Park & London Ctl	121.025, 133.700 abm Birmingham & 133.975
	Abm Birmingham & ROBIN	London Ctl 121.025, 133.700 & 134.425 above FL135
	Abm Birmingham & ROBIN	Manchester Ctl 124.200 & 126.650 below FL175
	ROBIN & Pole Hill	London Ctl 131.050, 134.425 above FL155
	ROBIN & Pole Hill	Manchester Ctl 124.200 & 126.650 below FL175
	Pole Hill & 54.30N	London Ctl 131.050 & 134.425 above FL155
	Pole Hill & 54.30N	Manchester Ctl 126.65 124.200 below FL175
	54.30N & GRICE	Scottish Ctl 135.675 (night) & 128.500 (day)
B5	Entire route	London Ctl 134.250, 127.950 & 133.525
B11	Within London FIR	London Ctl 134.450, 127.700 & 124.275
B29	Within London FIR	London Ctl 129.600 & 127.950

A/way	Sector	Control
B39	MALBY & RADNO	London Ctl 131.200
	RADNO & TOLKA	London Ctl 128.050
B53	Entire route	London Ctl 128.050 & 129.100 above FL155
	Entire route	Manchester Ctl 125.100 & 124.200 below FL175
B226	Entire route	Scottish Ctl 124.500
G1	West of Brecon	London Ctl 131.200
	Brecon & abm Woodley	London Ctl 132.800 & 131.200
	East of abm Woodley to London Ctl 134.900 & 127.100 FIR Boundary	
G27	North of 50.00N	London Ctl 127.700 & 124.275R1
R1	ORTAC to Ockham	London Ctl 134.450, 132.300, 127.700 & 124.275
	Ockham to FIR Boundary	London Ctl 129.600, 127.950, 133.450 & 133.520
R3	Wallasey & ROBIN	London Ctl 128.050, 129.100 & 134.425 above FL155
	Wallasey & ROBIN	Manchester Ctl 125.100 & 124.200 below FL175
R8	BRIPO to Southampton	London Ctl 132.600 & 124.275
	Southampton to Midhurst	London Ctl 134.450, 132.300, 127.700 & 124.270
	Midhurst & Dover	London Ctl 134.900, 127.100 & 124.275
R12	Entire route	London Ctl 129.600, 127.950, 133.450 & 133.520
R123	Entire route	London Ctl 129.600, 127.950, 133.450 & 133.520
R14	Within London FIR	London Ctl 131.200
R25	Entire route	London Ctl 127.700
R41	ORTAC & Southampton	London Ctl 134.450, 132.300, 127.700 & 124.275
	Southampton & abm Compton	London Ctl 132.800, 131.200, & 124.275
	Abm Compton & Westcott	London Ctl 133.700 & 121.025
	Entire route	London Ctl 134.450, 132.300, 127.700 & 124.275
R126	Within London FIR	London Ctl 129.600 & 127.940
R803	Entire route	London Ctl 127.700 & 124.275
W1	Daventry to Abm Barkway	London Ctl 121.025, 133.700, & 133.975
	Abm Barkway to 20nm N of Dover	London Ctl 129.600, 127.950, & 133.450
	20nm North of Dover to Dover	London Ctl 132.900 & 127.100
W923	Entire route	London Ctl 131.050, 129.100, & 134.425 above FL155
	Entire route	Manchester Ctl 126.650 & 124.200 below FL175
W934	Within London FIR	London Ctl 127.700 & 124.275

Lower ATS Advisory Routes

A/way	Sector	Control
A1D	60N 10W to Stornoway	Scottish Ctl 127.275
	Stornoway to Glasgow	Scottish Ctl 127.275
B1D	Within Scottish FIR	Scottish Ctl 131.300
G4D	Within London FIR	London Ctl 132.600
N552D	Entire route	Scottish Ctl 127.275
N562	Entire route	Scottish Ctl 127.275
N571D	Entire route	Scottish Ctl 127.275
R8D	Within London FIR	London Ctl 132.600
W2D	West of Fleetwood	London Ctl 128.050, 129.200, & 134.425 above FL155
	West of Fleetwood	Manchester Ctl 133.050 below FL175
	East of Fleetwood	London Ctl 131.050 & 134.425 above FL155
	East of Fleetwood	Manchester Ctl 126.650 & 124.200 below FL175
W3	South of Inverness	Scottish Ctl 124.500
	Between Inverness & Sumburgh	Scottish Ctl 131.300
W4D	Within Scottish FIR	Scottish Ctl 131.300

Lower ATS Advisory Routes

W5D	Within Scottish FIR	Scottish Ctl 131.300
W6D	Glasgow to Benbecula to	Scottish Ctl 127.275
	Stornoway to 05.00W to Inverness	
W910D	Entire route	Scottish Ctl 127 275
W911D	South of 54.30N	Scottish Ctl 128.500 & Border Radar on 132.900
	South of 54.30N	London Ctl 128.050, 129.100, & 134.425 above FL155
	South of 54.30N	Manchester Ctl 133.050 below FL175
W927D	West of North light	London Ctl 128.050, 129.100 & 134.425 above FL 155
	West of North light	Manchester Ctl 133.050 below FL175
	East of North light	London Ctl 128.050, 134.425, above FL155
	East of North light	Manchester Ctl 133.050 below FL175
W928D	Entire route	Scottish Ctl 135.675
W985D	Entire route	Scottish Ctl 127.275

Upper ATS allocations

UA1	North of 54.30N	Scottish Ctl 135.850
	Between 54.30N & abm Lichfield	London Ctl 131.050, 129.100 & 134.425
	Abm Lichfield & abm Woodley	London Ctl 133.700
	South of Woodley to UIR Boundary	London Ctl 127.700, 124.275 & 127.425
UA2	Machrihanish & 54.30N	Scottish Ctl 135.850
	54.30N & Trent	London Ctl 131.050, 129.100 & 134.425
	Trent & Lambourne	London Ctl 133.700 & 121.025
	South of Lambourne to London Ct	127.100, 132.450 UIR Boundary & 127.425
UA20	Entire route	London Ctl 127.100 & 127.425
UA25	GRICE to 54.30N	Scottish Ctl 135.850
	54.30N & South of Wallasey	London Ctl 128.050, 129.100 & 134.425
	South of Wallasey & S of Brecon	London Ctl 133.600
	South of Brecon to UIR	London Ctl 132.600, 131.050 Boundary 134.425
UA29	BAKUR & MERLY	London Ctl 133.600
	MERLY & SALCO	London Ctl 132.600
UA30	Entire route	London Ctl 127.100 & 127.425
UA34	Wallasey & TELBA	London Ctl 128.050 & 129.100
	TELBA & Abm Woodley	London Ctl 133.700
	Abm Woodley & UIR Boundary	London Ctl 127.700, 124.270 & 127.425
UA37	DANDI & GABAD	London Ctl 134.250, 128.125 & 133.525
	GABAD & Detling	London Ctl 129.600, 127.950, 133.525 & 127.400
UA47	Daventry & Woodley	London Ctl 133.700 & 121.025
	South of Woodley to London Ctl	127.700, 135.050, UIR Boundary 127.425
UA251	Pole Hill & TELBA	London Ctl 131.050 & 129.100
	TELBA & EXMOR	London Ctl 133.600
UB1	Liffey to Wallasey	London Ctl 128.050 & 134.425
	Wallasey to Ottringham	London Ctl 131.050 & 134.425
	East of Ottringham	London Ctl 134.250, 128.125 & 133.525
UB2	DALKY to Perth	Scottish Ctl 135.850, & 126.850
	Perth to KLONN	Scottish Ctl 124.050
UB3	Belfast to 05.00W	Scottish Ctl 135.85 & 126.850
	05.00W to 53.00N	London Ctl 128.050 & 121.025
	Brookmans Park & Dover	London Ctl 127.100, 134.900 & 127.425
UB4	FINDO & 54.30N	Scottish Ctl 135.850
	54.30N & ROBIN	London Ctl 131.050 & 134.425
	ROBIN & Brookmans Park	London Ctl 121.025 & 133.700
	South of Brookmans Park to London Ctl	127.100, 132.450 UIR Boundary 127.425
UB5	North of FAMBO	Scottish Ctl 135.850
	South of FAMBO	London Ctl 134.250, 128.125 & 133.525
UB10	Within London UIR	London Ctl 133.600
UB11	Within London UIR	London Ctl 134.450, 127 700, 124.275 & 127.400

Upper ATS allocations

UB29	Compton & Abm Brookmans Park	London Ctl 133.600 & 132.800
	East of Abm Brookmans Park	London Ctl 129.600, 127.950, 133.525
		to UIR Boundary & 127.400
UB39	Midhurst & RADNO	London Ctl 133.600 & 132.600
	RADNO & TOLKA	London Ctl 128.050 & 127.425
UB40	Entire route	London Ctl 133.600 & 132.600
UB105	Within London UIR	London Ctl 134.250,128.125 & 133.525
UG1	West of Abm Woodley	
	to London Ctl	133.600 & 132.800 & UIR Boundary
	East of Abm Woodley	
	to London Ctl	134.900, 127.100 & UIR Boundary 127.425
UG4	Within London UIR	London Ctl 132.600
UG11	Within Scottish UIR	Scottish Ctl 124.050
UG106	Within London UIR	London Ctl 134.900, 127.100 & 127.425
UH71	Sumburgh to LIRKI	Scottish Ctl 124.050 & 134.775
UH73	GRICE to Machrihanish	Scottish Ctl 135.850 & 126.850
UL1	West of abm Woodley	
	to London Ctl	133.600 & 132.800 & UIR Boundary
		East of abm Woodley
	to London Ctl	134.900, 127.100 & UIR Boundary 127.425
UL74	Entire route	London Ctl 134.250 & 128.125
UL7	North of SKATE	Scottish Ctl 124.050
	South of SKATE	London Ctl 134.250 & 128.125
UL722	Entire route	London Ctl 132.600 & 132.950
UN490	UIR Boundary to TAKAS	Brest Control 129.500
UN491	UIR Boundary to TAKAS	Brest Control 129.500
UN500	Entire route	London Ctl 132.600 & 132.950
UN508	UIR Boundary to TAKAS	Brest Control 129.500
UN510	RATKA to OMIMI	Shannon Ctl 135.600
UN520	OMIMI to UIR Boundary	Brest Ctl 129.5
UN549	Strumble to BAKUR	London Ctl 133.600 & 132 800
UN550	ERNAN to 55.00N 01.00W	Scottish Ctl 135.850 & 126.850
UN551	Belfast to 55.00N 01.00W	Scottish Ctl 135.850 & 126.850
UN552	TALLA to Machrihanish	
	to Scottish Ctl	135.850 & 126.850 & 55.00N 01.00W
UN560	ERNAN to 55.00N 01.00W	Scottish Ctl 135.850
UN561	Belfast to 55.00N 01.00W	Scottish Ctl 135.850 & 126.850
UN562	Machrihanish to 55.00N 01.00W	Scottish Ctl 135.850 & 126.850
UN563	Glasgow to 55.00N 01.00W	Scottish Ctl 135.850 & 126.850
UN564	GRICE to 55.00N 01.00W	Scottish Ctl 135.850 & 126.850
UN570	GRICE to 56.00N 01.00W	Scottish Ctl 135.850 & 126.850
UN571	Machrihanish to 57.00N 01.00W	Scottish Ctl 135.850 & 126.850
UN572	Tiree to 57.00N 01.00W	Scottish Ctl 135.850 & 126.850
UN580	Glasgow to Tiree to 57.20N	Scottish Ctl 135.850 & 126.850
		57.20N to 58.00N 01.00W Scottish Ctl
		124.050 & 134.775
UN581	Aberdeen to Benbecula	
	to Scottish Ctl	124.050 & 134.775 & 58.00N 01.00W
UN582	ASPIT to 57.20N	Scottish Ctl 135.850 & 126.850
	57.20N to Stornoway	Scottish Ctl 124.050 & 134.775
UN58	Sumburgh to Stornoway	to Scottish Ctl 124.050 & 134.775 & 58.00N
		01.00W
UN584	Sumburgh to 58.00N 01.00W	Scottish Ctl 124.050 & 134.775
UN590	MARGO to Glasgow & Benbecula	Scottish Ctl 135.850 & 126.850
	Benbecula to 59.00N 01.00W	Scottish Ctl 124.050 & 134.775
UN593	Sumburgh to 59.00N 01.00W	Scottish Ctl 124.050 & 134.775
UN601	TALLA to 57.20N	Scottish Ctl 135.850 & 126.850
	57.20N to Stornoway to	
	Scottish Ctl	124.050 & 134.775 & 60.00N to 01.00W
UN602	Glasgow to 57.20N	Scottish Ctl 135.850 & 126.850
	57.20N to RONAK to 60.00N	Scottish Ctl 124.050 & 134.775 & 01.00W
UN603	Sumburgh to 60.00N 01.00W	Scottish Ctl 124.050 & 134.775
UN610	Stornoway to 60.00N 01.00W	Scottish Ctl 124.050 & 134.775
UN611	BATSU to Aberdeen to	

Upper ATS allocations

	RONAK to 60.10N 01.00W	Scottish Ctl 124.050 & 134.775
UN612	Sumburgh to 60.10N 01.00W	Scottish Ctl 124.050 & 134.775
UN615	Glasgow to 57.20N	Scottish Ctl 135.850 & 126.850
	57.20N to Stornoway to MATIK	Scottish Ctl 124.050 & 134.775
UR1	ORTAC/Midhurst to Lambourne	London Ctl 134.450, 127.700
UR12	abm Lambourne	132.300 & 124.275
UR123	Lambourne to Clacton to	London Ctl 129.600, 127.950, 133.450, &
		UIR Boundary 133.525 & 127.425
UR3	Entire route	London Ctl 128.050 & 134.425
UR4	IOM to Pole Hill	London Ctl 128.050 & 134.425
	Pole Hill to Ottringham	London Ctl 131.050 & 134.425
	Ottringham to DANDI	London Ctl 134.250, 128.125 & 133.525
UR8	From Lands End to Southampton	London Ctl 132.600, 124.275, 134.450,
		132.300 & 127.700
	From Southampton to Midhurst	London Ctl 124.275 & 127.425
UR14	Within London UIR	London Ctl 132.600 & 133.600
UR23	Glasgow to SAB	Scottish Ctl 135.850
	SAB to GORDO	Scottish Ctl 124.050
UR24	ORIST to ASPEN	London Ctl 134.450, 132.300 & 127.700
UR25	Entire route	London Ctl 127.700 & 124.275
UR37	NORLA to Southampton	London Ctl 132.600 & 124.275
	Between Southampton	London Ctl 134.450, 127.700, & abm
		Midhurst 132.310 & 124.27
	Abm Midhurst to DONER	London Ctl 134.900, 127.100 124.275 &
		127.425
UR38	Newcastle to 57.15N	Scottish Ctl 135.850 & 126.850
	57.15N to Stornoway	Scottish Ctl 124.050 & 134.775
UR41	Between ORTAC & Southampton	London Ctl 134.450, 132.300 & 127.700
	Between Southampton	London Ctl 132.800 & 131.200 & abeam
		Woodley
	Between Abm Woodley	133.700, 121.025 & 127.425 & Westcott
UR84	ORTAC to Midhurst	London Ctl 132.300, 127.700 & 127.425
UR126	Entire route	London Ctl 129.600, 127.950 & 133.525
UR168	Lands End to CAVAL	London Ctl 132.600
UT7	From Lands End to NOTRO	London Ctl 132.600
	From NOTRO to ASKIL	Brest Ctl 129.500
UW1	Between Daventry & abm Barkway	London Ctl 121.025 & 133.700
	Between abm Barkway & Clacton	London Ctl 129.600, 127.950, 133.450 &
		133.500
UW2	Between Compton	London Ctl 133.600, 132.600 & Brookmans
		Park 127.425

Northern Radar Advisory Service area

North of W911D	Scottish Control 124.500
South of W911D	Pennine Radar 128.675

Hebrides Upper Control Area

South of a line 57.30N 10.00W	
& TIR & 65.36N & 00.41W	Scottish Ctl 135.850 & 126.850
North of a line 57.30N 10.00W	
& TIR & 65.36N & 00.41W	Scottish Ctl 124.050

Notes on civilian airport frequencies

Occasionally frequencies may be interchanged and, for instance, approach control will be handled by the tower. However, all the air-fields shown do have their main frequencies listed. A/G (Air/Ground stations) are for the most part communication stations available at smaller airfields. Pilots can call these facilities to obtain current weather information and the operator may well also warn of any other aircraft that are in the circuit. However, unlike an air traffic controller, the operator is not licensed to give the pilot landing instructions and it is up to the pilot to keep a look-out and make sure that he is not going to endanger any other aircraft during landing or take-off.

Military airport transmissions

Military airport transmissions use the same frequencies as civilian airports, but also have frequency allocations between 240 and 350MHz.

Special military allocations

Facility	Frequency (MHz)
Aces High Ops (N. Weald)	130.175
Boulmer Rescue	123.100
Dalcross Tower (range)	122.600
Distress (army)	40.0500
Distress (including beacons)	243.000
Donna Nook Range	123.050
Lee-on-Solent rescue	132.650
NATO emergency	243.000, 40.050
NATO low level manoeuvres	273.900
NATO SAR training	253.800
Neatishead (range warning)	123.100
SAR co-ordination air/sea	123.100
Spadeadam Range	122.100
Standard Mil Field frequency	122.100, 123.300
Wembury Range control	122.100

UK military danger area activity information service (DAAIS)

Aberdeen	120.400	Liverpool	119.850
Aberporth	122.150	Llanbedr Radar	122.500
Bentwaters	119.000	London Info	124.600
Border Info	134.850	London Info	134.700
Border Info	132.900	Lydd	120.700
Boscombe Down	126.700	Lyneham	123.400
Brawdy	124.400	Neatishead	123.100
Bristol	127.750	Newcastle	126.350
Brize Radar	134.300	Portland	124.150
Chivenor	130.200	St. Mawgan	126.500
Culdrose	134.050	Salisbury Plain	130.150
Dalcross Tower	122.600	Scottish Mil	124.900
Donna Nook	123.050	Scottish Mil	133.200
Edinburgh	121.200	Train Range	118.900
Farnborough	125.250	Waddington	127.350
Goodwood	122.450	Wembury Range	122.100
Leeming	132.400	West Freugh	130.050
Leuchars	126.500	Yeovilton	127.350

These stations provide information on the military training areas closest to them. Training ranges are used for a variety of purposes including gunnery on the ground, at sea and in the air. The airspace above the ranges is often closed to non-military aircraft and civilian pilots will call the above stations to determine whether or not they can fly through the areas.

MATZ stands for 'military aerodrome traffic zone' and civilian aircraft are not allowed in these areas without permission. Calls to obtain permission will be made on the MATZ frequency shown. As with the civilian listing, many of the frequencies are interchangeable and it is not unusual for, say, the MATZ Frequency to also be used for approach control or radar services.

Miscellaneous airband services

In addition to general airport approach, take off and landing services, there are a wide variety of other services. Aircraft need to be passed from one region to another and their use of designated airways needs to be controlled. When an aircraft is approaching London, for instance, it will have to change frequencies several times as it is handed from one sector to another. Crossing the borders of different countries also means a change of frequency to a new ground controller. Following an aircraft is easy as it is standard procedure in airband communication for the ground controller to tell the pilot which frequency to change to and for the pilot to repeat the frequency he has been given.

Emergency frequencies

In addition to being allocated for emergency communications use, frequencies 121.5MHz and 243.0MHz are also used for search and rescue beacons of three forms. The first is a small transmitter emitting a radio bleep. It is triggered automatically when a crash occurs, or may be switched on manually. The second type contains a voice transmitter. The third type also includes a receiver, so turning it onto a full, two-way communications transceiver. These beacons are either handheld or, in the case of SARBE versions, fitted to lifejackets. frequencies 156.0MHz and 156.8MHz (both marine frequencies) are used by search and rescue aircraft to communicate with lifeboats, etc.

Search and Rescue Primary frequencies

0.457MHz	Avalanche Rescue Beacons
2.18MHz	International Maritime distress, safety and calling
2.1875MHz	International maritime medium Frequency digital selective calling
2.272 MHz	Avalanche Rescue Beacons
4.340 MHz	NATO combined submarine distress.
8.364 MHz	Survival craft.
121.500 MHz	International aeronautical emergency Frequency
156.1250 MHz	Land SAR0
156.5250 MHz	International maritime VHF digital selective calling
156.8000 MHz	International maritime distress, calling and safety. FM
243.000 MHz	NATO combined Distress/Emergency.
406.0000-406.1000 MHz	
	Emergency Position-Indicating Radio Beacons (EPIRB) – maritime
	Emergency Locator Transmitters (ELT) – aviation
	Personal Locator Beacons (PLB) – multi-environmentSearch and Rescue Control and Scene of Search

3.023 MHz	RCC to ships and/or aircraft at scene of search.
5.680 MHz	RCC to ships and/or aircraft at scene of search.
8.364 MHz	International for intercommunication between survival craft, aircraft and ships.
86.3125 MHz	Land SAR. Used in Scotland as a feeder link for crossband rebroadcasting.
86.3250 MHz	Secondary land Search and Rescue (Low Band, shared)
123.1000 MHz	NATO/International combined scene of search and rescue
132.6500 MHz	Counter pollution
147.3500 MHz	Land SAR simplex or paired another suitable channel.
147.4750 MHz	Land SAR simplex or paired another suitable channel.
152.8500 MHz	Land SAR simplex or paired another suitable channel.
155.3500 MHz	Land SAR simplex or paired another suitable channel.
156.0000 MHz	Coastal and inshore search and rescue. FM
156.1750 MHz	Land SAR Team Working Channel. FM
156.2250 MHz	Land SAR Team Working Channel. FM
156.3000 MHz	Intership Frequency for use at scene of search. Also for RCC to ships at scene of search.
156.3750 MHz	May be used for SAR co-ordination by participating land, sea and air stations.
156.5000 MHz	May be used for SAR co-ordination by participating land, sea and air stations
156.6750 MHz	May be used for SAR co-ordination by participating land, sea and air stations
157.2000 MHz	Land SAR Ground to Air channel (abbreviated to G2A).
157.2750 MHz	Land SAR Team Working Channel.
158.6500 MHz	Land SAR Channel. FM
160.7250 MHz	Land SAR Team Working Channel.
160.7750 MHz	Land SAR Team Working Channel.
160.8250 MHz	Land SAR Team Working Channel.
161.8000 MHz	Land SAR Team Working Channel.
161.8750 MHz	Land SAR Team Working Channel.
168.3500 MHz	Scotland – Land SAR GPS Data Channel – paired with Channel 99
174.0375 MHz	Scotland – Land SAR GPS Data Channel – paired with Channel 98
244.6000 MHz	UK scene of SAR control. AM
252.8000 MHz	Military aviation scene of search. AM
282.8000 MHz	Combined scene of search and rescue. AM

RNLI Lifeboat (National Use, Simple Inversion Scrambling Used)

Ship	Coast	
162.14375	157.54375	Chan 1 (Crew)
162.15625	157.55625	Chan 2 (Launch)
159.1000	163.6000	Chan 3 (Private)
157.5500		Used by RNLI in Portsmouth (Marine Ch 31)

RNLI Beach Lifeguard Channels

161.2250		Newquay, Poole, and Weymouth
159.1000	163.6000	Newquay, Link to Marine Chan 0
162.14375	157.54375	Perranporth, Link to Marine Chan 0
162.15625	157.55625	Porthowan, Link to Marine Chan 0

Navigational aids

Between the frequencies 108MHz and 117.95MHz you will hear a variety of navigational aids. These are VHF omni-range beacons (VOR) and instrument landing systems (ILS). Some of these services are paired with navigational aids on other bands to give additional services such as distance measuring (DME). Combined VOR/DME services are called VORTACS & the TAC part being a shortening of TACAN which in turn stands for 'tactical navigation'.

VOLMETS

These are transmit-only stations providing constantly updated weather information for a variety of major airports.

VOLMETS providing weather information

Service	Frequency (MHz)
London VOLMET main London VOLMET north London VOLMET south Dublin VOLMET Scottish VOLMET	
135.375 126.600 128.600 127.000 125.725	

Miscellaneous airband allocations

Service	Frequency (MHz)
Air-to-Air	123.450
Air-to-Air (North Atlantic only)	131.800
Balloons (hot air)	129.900
Distress	121.500
Fisheries Protection surveillance	121.100 (North Sea)
	131.800 (S.W. approaches and Channel)
Fire vehicles	121.600
Gliders	130.100, 131.125, 130.400
Ground control	121.700, 121.800, 121.900
Hang gliders	129.900
Lighthouse helipads	129.700
Search & Rescue (SAR)	123.100

Airline and handling agent frequencies

Operator	Frequency (MHz)	Operator	Frequency (MHz)
Aceair	130.175	Gatwick Handling	130.650
Aer Lingus	131.500 131.750	Genavco	130.375
Air Atlantique	130.625	Hatair	123.650
Air Bridge Carriers	122.350	Iberia	131.950
Air Canada	131.450	Inflight	130.625
Air Foyle	131.775	Interflight	130.575
Air France	131.500	Iran Air	131.575
Air Hanson	130.375	Japanese Airlines	131.650
Air India	131.600	Jet centre	130.375
Air Jamaica	131.450	K.L.M.	131.650
Air Kilroe	122.350	Kuwait Airlines	131.500
Air Malta	131.650	Loganair	130.650
Air UK	129.750 131.750	Lufthansa	131.925
Alitalia	131.450	Luxair	131.550
Aurigny Aviation	122.350	Magec	123.650
British Aerospace	123.050	M.A.M.	129.700
Beauport Aviation	129.700	Manx Airlines	129.750
Britannia Airways	131.675	M.A.S.	131.575
British Air Ferries	130.625	McAlpine Aviation	123.650
British Airways	123.650 131.47	Monarch	131,525
	131.800 131.550	Nigerian Airways	131.775
	131.625 131.850	Northern	130.650
	131.900	Pakistan International	131.450
British Island Airways	129.750	Qantas	131,875
British Midland	129.750 131.575	Royal Jordanian	131.425
British West Indian	131.450	S.A.S.	131.700
Brymon	123.650	Sabena	131.475
Channel Express	130.600	Saudia	131.425
C.S.E. Oxford	129.700	Servisair	130.075 130.600
Connectair	130.175	Singapore Airlines	131.950
Cyprus Airways	131.775	Skycare	122.050 130.025
Danair	131.875	Spurnair	122.050
Delta	130.600	Swissair	131.700
Diamond Air	129.700	T.A.P.	131.750
Eastern Airlines	131.900	Thai-Inter	131.450
El Al	131.575	T.W.A.	131.600
Eurojet	131.875	Uni Avco	130.575
Execair	122.350	Veritair	129.900
Fields Heathrow	130.600	Wardair	129.700
Finnair 1	31.950		

Air Display Frequencies

All use AM unless stated
Commonly Used General Display CAA
Channels

121.175 Air-Ground
130.500 Air-Ground
130.675 Approach/Tower
134.550 General
132.900 Approach/Tower (Mildenhall)

Other general frequencies

118.375 Air-Ground Middle Wallop Helimeet
and Weston Helidays
118.600 Royal International Air Tattoo
119.450 Middle Wallop Helimeet - Tower
119.600 Mildenhall / Farnborough
123.375 Weston Helidays
123.775 Farnborough Approach
124.550 Farnborough Radar
128.075 Royal International Air Tattoo
130.825 Farnborough Ground
130.875 Approach
132.900 International Air Tattoo Fairford - Brize
Radar
134.025 Farnborough Tower
134.175 Farnborough Radar
134.500 Farnborough Radar
135.000 Tower
135.350 Farnborough Radar
135.750 Farnborough Radar

Air Display Teams

135.925 Green March (Morroccan) - back-up
channel
135.975 Green March (Morroccan) - primary
channel
136.175 Halcones (Chilean) 130.300 Aquilla
(Spanish display team)
130.500 Aquilla (Spanish display team)
130.550 Silver Swallows (Irish Air Corps)
130.725 Swallows (Belgian Air Force display
team)
143.100 Patrouile de Francaise
150.000 Team Apache
242.650 Patrouile de Francaise
243.450 Red Arrows
243.850 Patrouile de Francaise
252.500 Aquilla (Spanish display team)
255.100 Falcons (RAF parachutists)
279.600 Red Stars (Turkish F-5 team)
281.800 Dutch F-16 Air-Air and Air-Ground
288.850 Patrouile Suisse
300.025 French Jaguar Pair
307.800 Frecce Tricolori (Italian)
380.200 Blue Eagles
380.200 Army Air Corps Historic Flight
382.800 Blue Eagles
387.000 French Jaguar Pair FRNCH
JAGUAR PAIR

387.100 French Jaguar Pair FRENCH JAGUA
PAIR
440.450 NFM Frecce Tricolori - ground
crew commentator
445.3375 NFM Falcons throat microphones
air-ground
456.4625 NFM Jordanian Falcons ground crew
456.4625 NFM Israeli Air Force ground crew
464.250 NFM Red Devils drop zone air-
ground Chan 1
464.550 NFM Red Devils drop zone air-
ground Chan 2
462.925 NFM Red Devils drop zone air-ground
Chan 3

Unpowered Flight Frequencies

118.675 Paragliding, below 5000ft
122.475 Primary ballooning channel, also used
by hang gliders
129.825 Microlights
129.900 Parachutists plus some balloons
(ground to ground and recovery) and
gliders, also used for TV and filming,
129.975 Gliding, air to ground, e.g. Air Cadets
130.100 Glider channel, competition start and
finish, and training
130.125 Glider channel, training, cross-country
locations, and local flying
130.400 Glider channel, cloud flying and
cross-country location message relay

UK Air Band Control Frequencies

Chedburgh	118.475	London Control
Chedburgh	126.600	London Volmet
Chedburgh	127.100	London Control
Chedburgh	127.825	London Control
Chedburgh	128.125	London Control
Chedburgh	128.700	London Military
Chedburgh	132.600	London Control
Chedburgh	133.325	London Military East
Chedburgh	133.450	London Control
Chedburgh	233.800	London Military
Chedburgh	263.075	London Military East
Chedburgh	292.600	London Military East
Clee Hill	118.775	Manchester Control
Clee Hill	121.500	Distress
Clee Hill	124.750	London Control FIS
Clee Hill	126.875	London Control
Clee Hill	127.450	London Military
Clee Hill	128.250	London Military
Clee Hill	128.700	London Military
Clee Hill	131.050	London Control
Clee Hill	133.600	London Control
Clee Hill	133.900	London Military
Clee Hill	134.750	London Control
Clee Hill	135.150	London Military
Clee Hill	135.425	London Control
Clee Hill	244.375	London Military
Clee Hill	245.175	London Military
Clee Hill	247.275	London Military
Clee Hill	255.925	London Military
Clee Hill	257.275	London Military
Clee Hill	275.475	London Military

UK Air Band Control Frequencies

Clee Hill	278.025	London Military
Clee Hill	278.075	London Military
Clee Hill	283.525	London Military
Craigowl Hill	119.875	Scottish Control
Craigowl Hill	121.500	Distress
Craigowl Hill	124.500	Scottish Control
Craigowl Hill	134.475	Scottish Military
Craigowl Hill	259.175	Scottish Military
Craigowl Hill	268.925	Scottish Military
Craigowl Hill	292.675	Scottish Military
Craigowl Hill	358.875	Scottish Military
Daventry	127.100	London Control
Daventry	127.875	London Control
Daventry	128.600	London Volmet
Daventry	129.200	London Control
Daventry	131.125	London Control
Daventry	133.300	London Military
Daventry	245.000	London Military
Daventry	264.825	London Military
Daventry	270.000	London Military
Daventry	275.350	London Military
Davidstone Moor	121.500	Distress
Davidstone Moor	123.950	Shanwick Oceanic
Davidstone Moor	124.750	London Control FIS
Davidstone Moor	126.075	London Control
Davidstone Moor	127.650	Shanwick Oceanic
Davidstone Moor	127.700	London Control
Davidstone Moor	128.600	London Volmet
Davidstone Moor	129.375	London Control
Davidstone Moor	132.950	London Control
Davidstone Moor	133.300	London Military
Davidstone Moor	133.600	London Control
Davidstone Moor	135.150	London Military
Davidstone Moor	135.525	Shanwick Oceanic
Davidstone Moor	244.375	London Military
Davidstone Moor	245.175	London Military
Davidstone Moor	247.275	London Military
Davidstone Moor	255.925	London Military
Davidstone Moor	257.275	London Military
Davidstone Moor	258.975	London Military
Davidstone Moor	261.025	London Military
Davidstone Moor	262.975	London Military
Davidstone Moor	275.475	London Military
Davidstone Moor	278.025	London Military
Davidstone Moor	278.075	London Military
Davidstone Moor	283.525	London Military
Davidstone Moor	290.575	London Military
Grantham	120.025	London Control
Grantham	121.500	Distress
Grantham	123.950	Shanwick Oceanic
Grantham	124.600	London Control FIS
Grantham	127.650	Shanwick Oceanic
Grantham	133.525	London Control
Grantham	135.375	London Volmet
Grantham	135.525	Shanwick Oceanic
Grantham	254.825	London Military
Grantham	290.600	London Military East
Great Dun Fell	121.500	Distress
Great Dun Fell	125.475	London Control Fis
Great Dun Fell	126.600	London Volmet
Great Dun Fell	128.675	Manchester Control
Great Dun Fell	252.475	Scottish Military
Great Dun Fell	259.775	Scottish Military
Great Dun Fell	264.825	London Military

UK Air Band Control Frequencies

Great Dun Fell	299.975	London Military East
Greenford	121.500	Distress
Greenford	127.875	London Control
Greenford	128.250	London Military
Greenford	129.200	London Control
Greenford	243.000	Distress
Greenford	245.000	London Military
Greenford	275.350	London Military
High Buston	124.050	Scottish Military
High Buston	134.775	Scottish Control
High Buston	259.175	Scottish Military
High Buston	268.925	Scottish Military
Kelsal	118.775	Manchester Control
Kelsal	124.200	Manchester Control
Kelsal	125.100	Manchester Control
Kelsal	125.950	Manchester Control
Kelsal	126.650	Manchester Control
Kelsal	128.050	Manchester Control
Kelsal	129.100	London Control
Kelsal	131.050	London Control
Kelsal	133.400	Manchester Control
Kelsal	135.575	London Control
Kelsal	245.250	London Military
Kelsal	254.275	London Military
Kelsal	278.075	London Military
Lowther Hill	119.875	Scottish Control
Lowther Hill	124.500	Scottish Control
Lowther Hill	124.825	Scottish Control
Lowther Hill	126.300	Scottish Control
Lowther Hill	126.925	Scottish Control
Lowther Hill	127.275	Scottish Control
Lowther Hill	129.225	Scottish Control
Lowther Hill	134.300	Scottish Military
Lowther Hill	135.850	Scottish Control
Lowther Hill	249.475	Scottish Control
Lowther Hill	292.675	Scottish Control
Mangersta	121.500	Distress
Mangersta	123.950	Shanwick Oceanic
Mangersta	126.925	Scottish Control
Mangersta	127.275	Scottish Control
Mangersta	127.650	Shanwick Oceanic
Mangersta	132.725	Scottish Control
Mangersta	133.675	Scottish Control
Mangersta	134.300	Scottish Control
Mangersta	134.775	Scottish Control
Mangersta	135.525	Shanwick Oceanic
Mangersta	249.475	Scottish Military
Mount Gabriel	132.950	London Control
Pitful Head	133.875	Scottish Control
Pitful Head	231.625	Scottish Control
Pitful Head	258.000	Scottish Control
Preston	118.775	Manchester Control
Preston	124.200	Manchester Control
Preston	125.100	Manchester Control
Preston	125.950	Manchester Control
Preston	126.650	Manchester Control
Preston	127.450	London Military
Preston	128.050	Manchester Control
Preston	128.675	Manchester Control
Preston	129.100	London Control
Preston	131.125	London Control
Preston	133.050	Manchester Control
Preston	133.400	Manchester Control
Preston	134.475	Scottish Military

UK Air Band Control Frequencies

Preston	135.575	London Control
Preston	245.250	London Military
Preston	254.275	London Military
Prestwick Dundonald Hill	123.950	Shanwick Oceanic
Prestwick Dundonald Hill	124.825	Scottish Control
Prestwick Dundonald Hill	125.725	Scottish Volmet
Prestwick Dundonald Hill	126.300	Scottish Control
Prestwick Dundonald Hill	127.650	Shanwick Oceanic
Prestwick Dundonald Hill	135.525	Shanwick Oceanic
Prestwick Dundonald Hill	243.000	Distress
Reigate	128.425	London Control
Reigate	128.600	London Volmet
Reigate	129.425	London Control
Reigate	132.300	London Control
Reigate	132.600	London Control
Reigate	134.450	London Control
Reigate	135.325	London Control
Reigate	135.375	London Volmet
RHU Staffish	123.775	Scottish Control
RHU Staffish	125.675	Scottish Control
RHU Staffish	134.475	Scottish Military
RHU Staffish	135.850	Scottish Control
RHU Staffish	259.725	Scottish Military
RHU Staffish	268.575	Scottish Military
Rothwell	121.325	Scottish Control
Rothwell	125.275	Anglia Radar
Rothwell	126.775	London Control
Rothwell	128.925	Anglia Radar
Rothwell	131.225	London Military East
Rothwell	135.075	London Military East
Rothwell	135.275	London Military East
Rothwell	135.525	London Military East
Rothwell	135.925	London Military East
Rothwell	232.025	London Military East
Rothwell	248.775	London Military East
Rothwell	254.525	London Military East
Rothwell	275.675	London Military East
Rothwell	276.775	London Military East
Rothwell	277.775	London Military East
Rothwell	279.300	London Military East
Rothwell	280.575	London Military East
Rothwell	283.475	Anglia Radar
Rothwell	284.300	London Military East
Rothwell	291.775	London Military East
Rothwell	293.475	London Military East
Rothwell	299.975	London Military East
Rothwell	313.000	London Military East
Snaefell	121.500	Distress
Snaefell	123.775	Scottish Control
Snaefell	125.475	London Control Fis
Snaefell	126.875	London Control
Snaefell	133.050	Manchester Control
Snaefell	264.825	London Military
Stornaway	133.675	Scottish Control
Stornaway	231.625	Scottish Control
Stornaway	258.000	Scottish Control
Stornaway	259.725	Scottish Control

UK Air Band Control Frequencies

Stornaway	268.575	Scottish Control
Swingfield	118.475	London Control
Swingfield	120.025	London Control
Swingfield	121.500	Distress
Swingfield	127.825	London Control
Swingfield	128.425	London Control
Swingfield	132.450	London Control
Swingfield	133.450	London Control
Swingfield	134.900	London Control
Swingfield	135.425	London Control
Swingfield	136.600	London Control
Swingfield	299.800	London Military
Tiree	121.500	Distress
Tiree	125.675	Scottish Control
Tiree	127.275	Scottish Control
Tiree	129.225	Scottish Control
Tiree	132.725	Scottish Control
Tiree	249.475	Scottish Control
Trimigham	121.325	Scottish Control
Trimigham	121.500	Distress
Trimigham	124.600	London Control FIS
Trimigham	125.275	Anglia Radar
Trimigham	125.475	London Control FIS
Trimigham	126.775	London Control
Trimigham	128.125	London Control
Trimigham	131.225	London Military East
Trimigham	133.525	London Control
Trimigham	135.075	London Military East
Trimigham	135.625	London Military East
Trimigham	135.925	London Military East
Trimigham	138.275	London Military East
Trimigham	232.025	London Military East
Trimigham	248.775	London Military
Trimigham	254.825	London Military East
Trimigham	275.675	London Military East
Trimigham	276.775	London Military East
Trimigham	277.775	London Military East
Trimigham	279.300	London Military East
Trimigham	283.475	Anglia Radar
Trimigham	290.700	London Military East
Trimigham	291.775	London Military East
Trimigham	293.475	London Military East
Trimigham	299.975	London Military East
Ventnor	121.500	Distress
Ventnor	124.750	London Control FIS
Ventnor	126.075	London Control
Ventnor	128.600	London Volmet
Ventnor	129.425	London Control
Ventnor	132.300	London Control
Ventnor	135.050	London Control
Ventnor	135.375	London Volmet
Ventnor	136.325	London Control
Ventnor	136.600	London Control
Ventnor	244.375	London Military
Ventnor	251.225	London Military
Ventnor	261.025	London Military
Ventnor	268.975	London Military
Ventnor	290.700	London Military
Warlingham	124.800	London Control FIS
Warlingham	127.425	London Control
Warlingham	132.450	London Control
Warlingham	134.900	London Control
Warlingham	135.050	London Control
Warlingham	233.800	London Military

UK Air Band Control Frequencies		
Warlingham	251.225	London Military
Warlingham	275.475	London Military
Warlingham	291.075	London Military
Warlingham	299.800	London Military
Windy Head	119.875	Scottish Control
Windy Head	121.500	Distress
Windy Head	124.050	Scottish Military
Windy Head	124.500	Scottish Control
Windy Head	125.725	Scottish Volmet
Windy Head	126.925	Scottish Control
Windy Head	133.875	Scottish Control
Windy Head	134.300	Scottish Military
Windy Head	134.775	Scottish Control
Windy Head	249.475	Scottish Military

UK Air Band Control Frequencies		
Windy Head	252.475	Scottish Military
Windy Head	259.775	Scottish Military
Winstone	120.025	London Control
Winstone	126.075	London Control
Winstone	127.425	London Control
Winstone	127.700	London Control
Winstone	129.375	London Control
Winstone	133.900	London Military
Winstone	134.750	London Control
Winstone	261.025	London Military
Winstone	270.000	London Military
Winstone	290.575	London Military
Winstone	291.075	London Military

ACARS VHF Frequencies

131.725 MHz Primary channel in Europe
131.525 MHz European secondary
136.900 MHz Additional European Channel
131.475 MHz Air Canada company channel
131.550 MHz Primary Channel for USA and Canada, also Primary Channel for Australia
130.025 MHz Secondary channel for USA and Canada
129.125 MHz Additional channel for USA & Canada
130.450 MHz Additional channel for USA & Canada
131.125 MHz Additional channel for USA
131.450 MHz Primary channel for Japan

Marine band

This, like the VHF airband, is one of the few international bands; it is common to all ITU regions. It is channelized in that radio equipment made for marine VHF use does not usually have facilities to tune to a given frequency, instead it has a channel selector which goes from channel 1 to channel 88. A list of marine band channels and their transmission frequencies is given in the accompanying table. Channels are designated for specific uses in that some are for ship-to-shore use, others for ship-to-ship, and so forth. The method of operating on marine band is very different from airband. The use of GMDSS (Global Maritime Distress and Safety System) is employed in passenger-carrying and cargo vessels, where a non-speech DSC (Digital Selective Calling) transmission is initially made on channel 70, with a 'working' channel allocated for subsequent speech communication. For other vessels, such as yachts and similar pleasure craft that may not be fully GMDSS equipped, there is a common calling frequency for speech: channel 16, at 156.8MHz. When a ship wishes to call another station, even though the operator may know the channel that is used by that shore station, he will normally make first contact on channel 16. Once contact is made either on channel 16 or via DSC on channel 70, the ship and shore station will then move to a `working channel' - in most instances this will be the station's `prime' channel for general transmissions, or `link' channel for link calls (i.e., ship-to-shore telephone calls). Channels 0 and 67 are used by lifeboats and the coast guard. Some search and rescue aircraft also have the facility to work these channels. Many channels have two frequencies, this is to enable duplex operation. Because these are duplex transmissions, it is impossible for a scanner to simultaneously monitor both frequencies. Two scanners would have to be used, each tuned to one of the two frequencies, if the whole transmission was to be received.

Range

The useful range for marine VHF communications tends to be some what better than for the same type of frequencies and power levels used across land. Quite simply, there are few obstructions at sea and maximum ranges of 50-100 miles are not unusual. However, while signals from a ship may be quite strong at a coastal station, the signals may deteriorate even a mile or two inland.

International marine channels

Channel	Ship	Coast	Service
0	156.000		Coastguard/lifeboat
99	160.600		Coastguard/lifeboat
01	156.050	160.650	Port operations/link calls
02	156.100	160.700	Port operations/link calls
03	156.150	160.750	Port operations/link calls
04	156.200	160.800	Port operations/link calls
05	156.250	160.850	Port operations/link calls
06	156.300		Intership primary/search and rescue
07	156.350	160.950	Port operations/link calls
08	156.400		Intership
09	156.450		Intership
10	156.500		Intership/pollution control
11	156.550		Port operations
12	156.600		Port operations primary
13	156.650		Port operations
14	156.700		Port operations primary
15	156.750		Port operations
16	156.800		Calling channel
17	156.850		Port operations
18	156.900	161.500	Port operations
19	156.950	161.550	Port operations
20	157.000	161.600	Port operations
21	157.050	161.650	Port operations
22	157.100	161.700	Port operations
23	157.150	161.750	Link calls
24	157.200	161.800	Link calls
25	157.250	161.850	Link calls
26	157.300	161.900	Link calls
27	157.350	161.950	Link calls
28	157.400	161.200	Link calls
29	157.450	162.050	Private channel
30	157.500	162.100	Private channel
31	157.550	162.150	Private channel
32	157.600	162.200	Private channel

Channel	Ship	Coast	Service
33	157.650	162.250	Private channel
34	157.700	162.300	Private channel
35	157.750	162.350	Private channel
36	157.800	162.400	Private channel
37	157.850	162.450	Private channel (Marina use)
38	157.900	162.500	Private channel
39	157.950	162.550	Private channel
40	158.000	162.600	Private channel
41	158.050	162.650	Private channel
42	158.100	162.700	Private channel
43	158.150	162.750	Private channel
44	158.200	162.800	Private channel
45	158.250	162.850	Private channel
46	158.300	162.900	Private channel
47	158.350	162.950	Private channel
48	158.400	163.000	Private channel
49	158.450	163.050	Private channel
50	158.500	163.100	Private channel
51	158.550	163.150	Private channel
52	158.600	163.200	Private channel
53	158.650	163.250	Private channel
54	158.700	163.300	Private channel
55	158.750	163.350	Private channel
56	158.800	163.400	Private channel
60	156.025	160.625	Link calls
61	156.075	160.675	Link calls
62	156.125	160.725	Link calls
63	156.175	160.775	Link calls
64	156.225	160.825	Link calls
65	156.275	160.875	Link calls
66	156.325	160.925	Link calls
67	156.375		Intership/small yacht safety/ coastguard
68	156.425		Intership
69	156.475		Intership
70	156.525		Digital Selective Calling/Distress
71	156.575		Port operations
72	156.625		Intership
73	156.675		Intership/pollution control/ Coastguard
74	156.725		Ports/lock keepers/swing bridges
77	156.875		Intership
78	156.925	161.525	Port operations
79	156.975	161.575	Port operations
80	157.025	161.625	Port operations (Marina primary channel)
81	157.075	161.675	Port operations
82	157.125	161.725	Port operations
83	157.175	161.775	Port operations
84	157.225	161.825	Port operations
85	157.275	161.875	Port operations
86	157.325	161.925	Link calls
87	157.375	161.975	Link calls
88	157.425	162.025	Link calls
M1	157.850		Marinas
M2	161.675		Marinas & yacht clubs

On-board-ship UHF handset frequencies (FM-split frequency simplex)

CH1 457.525 paired with 467.525
CH2 457.550 paired with 467.550
CH3 457.575 paired with 467.675

General marine services and channel allocations

Service Channels

Ship-to-ship Port operations (simplex) Port operations (duplex) Public correspondence (link calls)
6, 8, 9, 10, 13, 15, 17, 67, 68, 70, 72, 75, 76, 77, 78 9, 10, 11, 12, 13, 14, 15, 17, 67, 69, 71, 73, 74 1, 2, 3, 4, 5, 7, 18, 19, 20, 21, 22, 60, 61, 62, 63, 64, 65, 66, 78, 79, 80, 81, 82, 84 1, 2, 3, 4, 5, 7, 23, 24, 25, 26, 27, 28, 60, 61, 62, 63, 64, 65, 66, 82, 83, 84, 85, 86, 87, 88

Shore stations

All stations transmit local area navigation warnings (beacons out of action, hazardous floating objects, etc). Most, but not all, transmit local area weather forecasts and storm warnings.

British and Irish Coastal Stations

Traffic	Nav lists	Wx Warn	Gale warn		
Cullercoats					
Chs: 16 26					
0103	1303	0233			
	0303	1503	0633	0703	0303
0503	1703	1033	0903		
0703	1903	1433	1503		
0903	2103	1833	1903	2103	
1103	2303	2233			
Hebrides					
Chs: 16 26					
0103	1303	0203			
	0303	1503	0603	0703	0303
0503	1703	1003	0903		
0703	1903	1403	1503		
0903	2103	1803	1903	2103	
1103	2303	2203			
Humber					
Chs: 16 24 26 85					
0103	1503	0133	0303		
0303	1703	0533	0733	0903	
0503	1903	0933	1503		
0903	2103	1333	2103		
1103	2303	1733	1933		
1303	2133				
Ilfracombe					
Chs: 16 05 07					
0133	1533	0233	0303		
0533	1733	0633	0833	0933	
0733	1933	1033	1503		
0933	2133	1433	2103		
1133	2333	1833	2033		
1333	2233				
Jersey					
Chs: 16 82 25 67					
After wx	0433	0645	as req +		
0645	0745				
0745	1245				
0833	1845	0307			
1245	2245	0907			
1633	1845	1507			
2033	2245	2107			
Land's End					
Chs: 16 27 88 85 64					
0103	1503	0233	0303		

Traffic	Nav lists	Wx Warn	Gale warn	
0303	1703	0633	0733	0903
0503	1903	1033	1503	
0903	2103	1433	2103	
1103	2303	1833	1933	
1303	2233			

Malin Head
Chs: 16 23 67 85

0103	1503	0033		
0503	1703	0433		
0903	1903	0833		
1103	2103	1233		
1303	2303	1633		

Niton
Chs: 16 04 28 81 85 64 87

0103	1503	0233	0303	
0303	1703	0633	0733	0903
0503	1903	1033	1503	
0903	2103	1433	2103	
1103	2303	1833	1933	
1303	2233			

North Foreland
Chs: 16 05 26 66 65

0103	1503	0233	0303	
0303	1703	0633	0733	0903
0503	1903	1033	1503	
0903	2103	1433	2103	
1103	2303	1833	1933	
1303	2233			

Portpatrick
Chs: 16 27

0103	1303	0203		
0303	1503	0603	0733	0303
0503	1703	1003	0903	
0703	1903	1403	1503	
0903	2103	1803	1903	2103
1103	2303	2203		

St Peter Port
Chs: 16 12 62 78

After	0133			
Nav	0533			
Warns	0933			
1333				
1733				
2133				

Shetland
Chs: 16 27

0103	1303	0233		
0303	1503	0633	0703	0303

Remote control from Wick

0503	1703	1033	0903	
0703	1903	1433	1503	
0903	2103	1833	1903	2103
1103	2303	2233		

Stonehaven
Chs: 16 26

0103	1303	0233		
0303	1503	0633	0733	0303
0503	1703	1033	0903	
0703	1903	1433	1503	
0903	2103	1833	1933	2103
1103	2303	2233		

Valentia
Chs: 16 24 28 67

0333	1533	0233	0033	

SCANNERS

Traffic	Nav lists	Wx Warn	Gale warn
0733	1733	0633	
0933	1933	1033	0633
1333	2333	1833	1233
2233	2033	1833	

Anglesey	Chs: 16 26 28 61 remote control by Portpatrick
Bacton	Chs: 16 07 63 64 03 remote control by Humber
Bantry	Chs: 16 23 67 85 remote control by Valentia
Belmullet	Chs: 16 67 83 remote control by Malin Head
Buchan	Chs: 16 25 87 remote control by Stonehaven
Cardigan Bay	Chs: 16 03 remote control by Portpatrick
Celtic	Chs: 16 24 remote control by Ilfracombe
Clyde	Chs: 16 26 remote control by Portpatrick
Collafirth	Chs: 16 24 remote control by Wick
Cork	Chs: 16 26 67 remote control by Valentia
Cromarty	Chs: 16 28 84 remote control by Wick
Dublin	Chs: 16 83 remote control by Malin Head
Forth	Chs: 16 24 62 remote control by Stonehaven
Glen Head	Chs: 16 24 67 remote control by Malin Head
Grimsby	Chs: 16 04 27 remote control by Humber
Hastings	Chs: 16 07 63 remote control by North Foreland
Islay	Chs: 16 25 60 remote control by Portpatrick
Lewis	Chs: 16 05 remote control by Stonehaven
Mine Head	Chs: 16 67 83 remote control by Valentia
Morcambe Bay	Chs: 16 04 82 remote control by Portpatrick
Orfordness	Chs: 16 62 82 remote control by North Foreland
Orkney	Chs: 16 26 remote control by Wick
Pendennis	Chs: 16 62 66 remote control by Land's End
Rosslare	Chs: 16 23 67 remote control by Valentia
Shannon	Chs: 16 24 28 67 remote control by Valentia
Skye	Chs: 16 24 remote control by Stonehaven
Start Point	Chs: 16 26 65 60 remote control by Land's End
Thames	Chs: 16 02 83 remote control by North Foreland
Weymouth Bay	Chs: 16 05 remote control by Niton
Whitby	Chs: 16 25 28 remote control by Cullercoats
Wick	See Shetland

*Remote stations broadcast at the same time as their control stations.

Ports, harbours and marinas in Great Britain

Port/Harbour Channels

Aberdeen (Aberdeen)
Aberdeen Radio 10 11 12 13 16
Aberdovey (Gwynedd)
Aberdovey Harbour 12 16
Abersoch (Gwynedd)
South Caernarfon Yacht Club 37 80
Aberystwyth (Dyfed)
Harbour Control 14 16
Alderney see Braye
Amble (Northumberland)
Amble Harbour 14 16
Amble Braid Marina 37 80
Appledore (Devon, see River Taw)
Ardrishaig (Argyll)
Harbour 16 74
Avonmouth (Avon) see Bristol
Barmouth (Gwynedd)
Barmouth Harbour 10 16
Barry (South Glamorgan)
Barry Radio 10 12
Beaucette (Guernsey)
Marina 16 37 80

Port/Harbour Channels

Bembridge (Isle of Wight)
Bembridge Marina 16 37 80
Berwick-on-Tweed (Northumberland)
Pilots and Harbourmaster 12 16
Blackwater River (Essex)
Bradwell & Tollesbury Marinas 37 80
Blyth (Northumberland)
Blythe Harbour Control 12 16
Boston (Lincolnshire)
Boston Dock 12
Grand Sluice 74
Braye (Alderney)
Alderney Radio 16 74
Mainbryce Marina (summer only) 37 80
Bridlington (Humberside)
Bridlington Harbour 12 14 16
Bridport (Dorset)
Bridport Radio 11 12 14 16
Brighton (East Sussex)
Brighton Control 11 16 68
Marina 37 80
Bristol (Avon)

Port/Harbour Channels

Avonmouth Radio 12
South Pier & Royal Edward Dock 12
Port Operations 09 11 14 16
Pilots 06 08 09 12 14
Royal Portbury, Portishead & City Docks 12 14 16
Floating Harbour 16 73
Newport 09 11 16
Brixham (Devon)
Harbour 14 16
Pilots 09 10 13 16
Brixham Coastguard 10 16 67 73
Bude (Cornwall)
Bude Radio
12 16
Burghead (Moray)
Harbour 14 16
Burnham-on-Crouch (Essex)
Essex and West Wick Marinas 37 80
Burnham-on-Sea (Somerset)
Harbour Master and Pilots 08 16
Watchet 09 12 14 16
Marina 37 80
Caernarfon (Gwynedd)
Caernarfon Radio (day only) 12 14 16
Caledonian Canal (Inverness) see also Inverness
All locks 74
Campbelltown (Argyll)
Harbour 12 14 16
Cardiff (South Glamorgan)
Docks 11 14 16
Marina 37 80
Cattewater Harbour (Devon) see Plymouth-Devonport
Charlestown (Cornwall) see Fowey
Chichester (West Sussex)
Harbour 14 16
Marinas 37 80
Colchester (Essex)
Colchester Harbour Radio 11 14 16
Conwy (Gwnedd)
Conway 06 08 12 14 16 72
Llanddulas 14 16
Cruising Club 37 80
Cowes (Isle of Wight)
Cowes 06 11 16
Island Harbour and marinas 37 80
Chain Ferry 10
Craobh Haven/Loch Shuna (Argyll)
Marina 16 37 80
Crinan (Argyll)
Harbour 16 74
Cromarty Firth
see Inverness
Dartmouth (Devon)
Harbour/Pilots 14
Dart Marina & Sailing Centre 37 80
Kingswear Marina ('Marina Four') 37 80
Fuel barge 16
Water taxi 16 37 80
Devonport (Devon) see Plymouth-Devonport

Port/Harbour Channels

Douglas (Isle of Man)
Douglas Harbour 12 16
Dover (Kent)
Dover Port Control 12 74
Channel Navigation Information Service 11 16 67 69 80
Information broadcasts (H+40) 11
Dundee/River Tay (Fife/Angus)
Dundee Harbour Radio 10 11 12 13 14 16
Perth Harbour 09
East Loch Tarbert (Argyll)
Harbour 16
Exeter (Devon)
Harbour 06 12 16
Pilots 09 12 14 16
Eyemouth (Berwick)
No regular watch 12 16
Falmouth (Cornwall) see also River Fal
Falmouth Harbour Radio 11 16
Felixstowe (Suffolk) see Harwich
Firth of Forth (Lothian Fife)
North Queensferry Naval Station 13 16 71
Rosyth Naval Base ('QHM') 13
Forth Navigation Service 12 16 20 71
Grangemouth Docks 14 16
Port Edgar Marina 37 80
BP Grangemouth Terminal 14 16 19
Braefoot Terminal 15 16 69 73
Hound Point Terminal 09 10 12 16 19
Fishguard (Dyfed)
Fishguard Radio 14 16
Marina 37 80
Fleetwood (Lancashire)
Fleetwood Harbour Control 11 12 16
Fleetwood Docks 12 16
Ramsden Docks 12 16
Folkestone (Kent)
Harbour 16 22
Pilot Station 09
Fowey (Cornwall)
Fowey Harbour Radio 12 16
Pilots 09
Water taxi 06
Boat Marshall Patrol 12 16
Charlestown 14 16
Par Port Radio 12 16
Fraserburgh (Aberdeen)
Harbour 12 16
Glasson Dock (Lancashire)
Glasson Radio 08 16
Marina M
Gorey (Jersey)
Gorey Harbour (summer only) 74
Gravesend (Essex)
Gravesend Radio 12 16
Great Yarmouth (Norfolk)
Yarmouth Radio 09 11 12 16
Breydon Bridge 12
Greenock (Renfrew)
Clydeport Estuary Radio 12 14 16
Dunoon Pier 12 16 31
Grimsby (Humberside) see also River Humber
Grimsby (Royal Dock) 09 16 18

Port/Harbour Channels

Guernsey see St Peter Port, St Sampsons & Beaucette
Hamble (Hampshire)
Hamble Harbour Radio 16 68
Marinas 37 80
Hartlepool (Cleveland)
Hartlepool Dock Radio 11 12 16
Harwich (Essex)
Harwich Harbour Control 11 14 16 71
Harbour Board patrol launch 11
Shotley Poin Marina 37 80
Havengore (Essex) see Shoeburyness
Helensburgh (Dumbarton)
Rhu Marina 37 80
Faslane (nuclear sub base) Patrol Boats 16
Helford River (Cornwall)
Helford River SC & Gweek Quay Marina 37 80
Heysham (Lancashire)
Heysham 14 16
Holyhead (Gwynedd)
Holyhead Radio 14 16
Anglesey Marine Terminal 10 12 16 19
Ilfracombe (Devon)
Ilfracombe Harbour (summer only) 12 16 37 80
Immingham (Humberside) see also River Humber
Immingham Docks 09 16 22
Inverkip (Renfrew)
Kip Marina 37 80
Inverness (Inverness)
Inverness Harbour Office 06 12 14 16
Inverness Boat Centre & Caley Marina 37 80
Cromarty Firth Port Control 06 08 11 12 13 14 16
Clachnaharry Sea Lock & Caledonian Canal 74
Ipswich (Suffolk)
Ipswich Port Radio 12 14 16
Marinas and yacht harbour 37 80
Isle of Man
Radio/Landline link to Liverpool 12 16
Isles of Scilly
Land's End Radio 64
St Mary's Harbour 14 16
Jersey
see St Helier & Gorey
King's Lynn (Norfolk)
King's Lynn Radio 11 1416
Docks 11 14 16
Wisbech 09 14 16
Kirkcudbright (Kirkcudbrightshire)
Harbour 12 16
Kirkwall (Orkney Islands)
Kirkwall Radio 12 16
Orkney Harbour Radio 09 11 16 20
Langstone (Hampshire)
Langstone 12 16
Marina 37 80
Largs (Ayr)
Yacht Haven 37 80
Lerwick (Shetland Islands)
Lerwick 11 12 16
Sullom Voe 12 14 16 19 20
Scalloway 12 16

Port/Harbour Channels

Balta Sound (no regular watch) 16 20
Littlehampton (West Sussex)
Littlehampton 14 16
Marina 37 80
Liverpool (Merseyside)
Mersey Radio 09 12 16 18 19 22
Alfred & Gladstone Docks 05
Tranmere Stages 09
Gartson & Waterloo Docks 20
Langton Dock 21
Eastham Locks (Manchester Ship Canal) 07 14
Latchford Locks (Manchester Ship Canal) 14 20
Weaver Navigation & Weston Point 74
Marsh, Dutton & Saltisford Locks 74
Anderton Depot 74
Loch Craignish (Argyll)
Yachting Centre 16 37 80
Loch Maddy (North Uinst)
Harbour 12 16
Loch Melfort (Argyll)
Camus Marine 16 37 80
London see River Thames
Looe (Cornwall)
Only occasional watch 16
Lossiemouth (Moray)
Lossiemouth Radio 12 16
Lowestoft (Suffolk)
Harbour 14 16
Pilots 14
Lyme Regis (Dorset)
Lyme Regis Harbour Radio 14 16
Lymington (Hampshire)
Marinas 37 80
Lulworth Gunnery Range (Dorset)
Range Safety Boats 08
Portland Naval Base 13 14
Portland Coastguard 67
Macduff (Banff)
Harbour 12 16
Mallaig (Inverness)
Mallaig Harbour Radio 09 16
Manchester (Lancashire)
Ship canal 14 16
Barton & Irlham Docks & Mode Wheel Lock 14 18
Stanlow Oil Docks & Latchford Lock 14 20
Eastham Lock 07 14
Tugs inbound 08
Tugs outbound 10
Weaver Navigation Service 14 71 73
Maryport (Cumbria)
Maryport Harbour (occasional watch) 12 16
Methil (Fife)
Methil Radio 14 16
Mevagissey (Cornwall)
Harbour 16 56
Milford Haven (Dyfed)
Milford Haven Radio 09 10 11 12 14 16 67
Patrol & Pilot launches 06 08 11 12 14 16 67
Milford Docks 09 12 14 16
Amoco/Gulf Terminals 14 16 18
Esso Terminal 14 16 19
Texaco Terminal 14 16 21

Port/Harbour Channels

Marina and Yacht Station 37 80
Minehead (Somerset)
Minehead Radio (occasional watch only) 12 14 16
Montrose (Angus)
Montrose Radio 12 16
Newhaven (East Sussex)
Harbour 12 16
Marina 37 80
Newlyn (Cornwall)
Newlyn Harbour 12 16
Pilots 09 12 16
Newquay (Cornwall)
Newquay Radio 14 16
North Shields (Tyne and Wear) see River Tyne
Oban (Argyll)
Coastguard 16
Padstow (Cornwall)
Padstow Radio 16 14
Par Port Radio (Cornwall) see Fowey
Peel (Isle of Man)
Peel 12 16
Penzance (Cornwall)
Harbour/Pilots 09 12 16
Peterhead (Aberdeen)
Peterhead Radio 09 11 14 16
Plymouth - Devonport (Devon)
Long Room Port Control 08 12 14 16
Mill Bay Docks 12 14 16
Sutton Harbour Radio 12 16 37 80
Marinas, Yacht harbour & Clubs 37 80
Cattewater Harbour Office 12 16
Poole (Dorset)
Poole Harbour Control 14 16
Pilots 06 09 14 16
Salterns Marina ('Gulliver Base') 37 80
Cobb's Quay 37 80
Porthmadog (Caernarfon)
Harbour Master 12 16
Madoc Yacht Club 16 37
Portland (Dorset)
Portland Naval Station 13 14
Portland Coastguard 16 67 69
Port of London see River Thames
Portpatrick (Wigtown)
Portpatrick Coast Radio Station 16 27
Stranraer 14 16
Portree (Skye)
Port (no regular watch) 08 16
Port Saint Mary (Isle of Man)
Port Saint Mary Harbour 12 16
Portsmouth (Hampshire) see also Solent
Portsmouth Harbour Radio 11 13
Queens Harbour Master 11
Portsmouth Naval 13
Marina ('Camper Base' & yacht harbour 37 80
Fort Gilkicker 16
Ramsey (Isle of Man)
Ramsey 12 16
Ramsgate (Kent)
Harbour & Marina 14 16
River Avon see Bristol
River Deben (Suffolk)
Tide Mill Yacht Harbour 37 80

Port/Harbour Channels

Pilot 08
River Exe (Devon)
Exeter 06 12 16
River Fal (Cornwall)
Falmouth Harbour Radio 11 16
Pilots 06 08 09 10 11 12 14 16
Coastguard 10 16 67 73
Customs Launch 06 09 12 16
Port Health 06 12 16
Yacht harbour, marina and club 37 80
River Humber (Humberside)
Humber Vessel Traffic Service 12 16
Grimsby (Royal Dock) 09 16 18
Immingham Docks 09 16 22
River Hull Port Operations ('Drypool Radio') 06 14 16
Tetney Oil Terminal 16 19
Goole Docks 14 16
Booth Ferry Bridge 12 16
Selby Bridges 09 12 16
Marinas and yacht harbour 37 80
River Medway (Kent) see also The Swale
Medway Radio 09 11 16 22 74
Marinas 37 80
River Ore (Suffolk)
Orford 16 67
River Orwell (Suffolk) see Ipswich
River Taw (Devon)
Pilots 06 09 12 16
River Tees (Cleveland)
Tees Harbour Radio 08 11 12 14 16 22
River Thames (London)
Woolwich Radio 14 16 22
Gravesend Radio 12 14 16 18 20
Thames Radio 02
Thames Patrol 06 12 14 16
Thames Barrier 14
Thames Navigation Service 12
St Katherines Yacht Haven 37 80
Chelsea Harbour Marina 14 16 37
Brentford Dock Marina 14 16
North Foreland 26
Hastings 07
Orfordness Radio 62
Shellhaven 16 19
River Tyne (Tyne and Wear)
Tyne Harbour Radio 11 12 14 16
Rona Naval Base (Rona)
Base 16
Rothesay (Isle of Bute)
Harbour 12 16
Rye (East Sussex)
Harbour 14 16
Saint Helier (Jersey)
Saint Helier Port Control 14
Jersey Radio 16 25 82
Lifeboat 00 16 14
Saint Kilda (Saint Kilda Island)
Kilda Radio 08 16
Saint Peter Port (Guernsey)
Port Control 12 16 21
Lifeboat 00 12 16
Saint Sampsons (Guernsey)
Harbour 12 16

115

SCANNERS

Port/Harbour Channels

Salcombe (Devon)
Salcombe Harbour 14
ICC Clubhouse & Floating HQ 37 80
Fuel barge 06
Water taxi 14
Scarborough (North Yorkshire)
Scarborough Lighthouse 12 14 16
Scrabster (Caithness)
Harbour 12 16
Seaham (Durham)
Seaham Harbour 06 12 16
Sharpness (Gloucestershire)
Sharpness Control 14 16
Bridges 74
Sheerness (Kent) see River Medway
Shetland Islands
Shetland Radio 16 27
Collafirth Radio 16 24
Shoeburyness (Essex)
Gunnery Range Operations Officer 16
Gravesend Radio 12
Shoreham (West Sussex)
Marinas 37 80
Solent (Hampshire)
Solent Coastguard 00 06 10 16 67 73
Southampton Port Radio 12 14 16 18 20 22
Pilots 06 08 09 10 12 14 16 18
Queen's Harbour Master (Portsmouth) 11 13
Commercial Harbour Master (Portsmouth) 11
Pilots 09
Ships, tugs, & berthing 71 74
BP Terminal 06 16 18
Esso Terminal 14 16 18
Southampton (Hampshire) see also Solent
Vessel Traffic Services 12 14 16
Harbour Patrol 10 12 14 16 18 22 71 74
Marinas 37 80
Southwold (Suffolk)
Southwold Port Radio 12 16
Pilots 09 12
Stornoway (Outer Hebrides-Lewis)
Harbour 12 16
Stromness (Orkney Islands)
Stromness Radio 12 16
Sullom Voe see Lerwick
Sunderland (Tyne and Wear)
Sunderland Docks 14 16
Sutton Harbour (Devon) see Plymouth-Devonport
Swansea (West Glamorgan)

Port/Harbour Channels

Swansea Docks Radio 14 16
Marina 37 80
Teignmouth (Devon)
Harbour/Pilots 12 16
Tenby (Dyfed)
Listening watch days only 16
The Swale (Kent)
Medway Radio 09 11 16 22 74
Kingsferry Bridge 10
Torquay (Devon)
Harbour 14 16
Troon (Ayr)
Marina 37 80
Androssan 12 14 16
Girvan 12 16
Ullapool (Ross and Cromarty)
Port 12 16
Watchet (Somerset)
Port/Pilots 09 12 14 16
Wells-next-the-Sea (Norfolk)
Wells Radio 06 08 12 16
Weymouth (Dorset)
Harbour 12 16
Pilots 09 16
Whitby (North Yorkshire)
Whitby Harbour 11 12 16
Whitby Bridge 06 11 16
Whitehills (Banff)
Whitehills Harbour Radio 09 16
Whitstable (Kent)
Harbour 09 12 16
Wick (Caithness)
Harbour 14 16
Wisbech (Cambridgeshire)
Wisbech Cut 09 14 12
Workington (Cumbria)
Workington Docks 14 16

Principal simplex services' channel allocations

Service	Channel
Calling and distress	16
Port operations (prime)	12
Port operations (alternative)	14
Small yacht safety	67
Marinas	M/M2/80
Inter-ship (prime)	06
Inter-ship (alternative)	08

Rail Communication

Band III Radio Telephone and UHF Cab Secure Radio

There are currently two sets of frequencies in use for train communication. The first, which is an 'area wide' system, operates on Band III (174-225MHz) and is commonly called 'Radio telephone' by rail users.

116

The second system operates on 450MHz UHF and is called 'Cab Radio', or to give it its full title "Cab Secure Radio" or "CSR" for short. Each fulfils a slightly different purpose.

Band III

This is a trunked radio system, using the MPT1327 system with base stations located in an 'area' configuration, in the same was as amateur and PMR speech repeaters. These each cover a given geographical area, albeit with some 'dead spots' in cuttings and tunnels, and if you've ever used a cellphone on a train, you'll know what I mean! In use, the driver keys his 'area code' into the radio, and the radio then automatically locates and locks onto the 'control channel', which sends periodic data out with area identification, call 'ahoys' to train radios etc. The operation is very similar to a telephone expect that it's simplex mode, and to make a call the driver enters an area number, just like a telephone 'STD' code, followed by the rail network telephone number. Each loco is allocated a four-digit number, and the reverse can also be accomplished from a control station where a loco can be called. Conversations usually follow a telephone format, but with an 'over' at the end of each transmission, and calls are limited to three minutes maximum. Outside calls can be made to PSTN (e.g. BT) lines, and an often-used call is that to the 'speaking clock' for the driver to check his timings! On suitably-equipped radios a paging message can be sent from control and displayed on the train-based radio's display, using the data messaging capabilities of MPT1327. The Band III base antennas can be identified as tower-mounted vertical folded dipoles, these are around three quarters of a metre long from tip to tip.

UHF Cab Radio

This is a further voice and data system, principally between signalmen and drivers, with the carefully-planned radio coverage concentrated along the length of railway lines. A large number of small lattice masts with directional UHF yagis have appeared over the last couple of years for this. You'll typically see at least couple of yagis on each mast, often mounted 'back to back' and beamed along the direction of the track. This way, coverage through cuttings and tunnels is vastly improved, and coverage of non-railway areas reduced. There are at least five of these masts within a few miles of my house, the nearest just a few hundred metres away on a single-track freight line.

The Cab Radio system allows direct data and voice communication between the signalman and the driver, so that if an incident occurs on the track or any other part of the railway system any needed details can be shared with each. It also gives the signalman 'live' data on the train's position including such information as Stock Number and Running Number. The system can also connect telephone calls via the signalman to the driver and vice versa. The signalman can even make 'public address' announcement to passengers on the train either through the driver or ''straight through';, useful if there's an accident in the cab. Various CTCSS tones are used on a geographical basis to give the best use of channels.

The accompanying frequency listings give the channels used on both Band III Radio Telephone and UHF Cab Radio.

Rail Band III Channels

TX Freq	RX Freq	Chan.Status	TX Freq	RX Freq	Chan.Status
204.8500	196.8500	Traffic	205.6750	197.6750	Emergency
204.9000	196.9000	Traffic	205.7000	197.7000	Traffic
204.9500	196.9500	Control	205.7250	197.7250	Control
205.0000	197.0000	Control	205.7500	197.7500	Traffic
205.0500	197.0500	Traffic	205.8000	197.8000	Traffic
205.1000	197.1000	Traffic	205.8375	197.8375	Control
205.1500	197.1500	Traffic	205.8500	197.8500	Traffic
205.2000	197.2000	Early Warning	205.9000	197.9000	Traffic
205.2500	197.2500	Early Warning	205.9500	197.9500	Traffic
205.3000	197.3000	Early Warning	206.0000	198.0000	Traffic
205.3500	197.3500	Traffic	206.1000	198.1000	Control
205.4000	197.4000	Traffic	206.1500	198.1500	Traffic
205.6000	197.6000	Traffic	206.2500	198.2500	Control
205.6500	197.6500	Traffic	206.3000	198.3000	Traffic

Cab Secure radio channels

Chan	Base TX	Train TX	Chan	Base TX	Train TX
1	454.84375	448.34375	12	454.98125	448.48125
2	454.85625	-	13	455.5125	449.0125
3	454.86875	448.36875	14	455.5875	449.0875
4	454.88125	448.38125	15	455.6625	449.1625
5	454.89375	448.39375	16	455.6750	449.1750
6	454.90625	448.40625	17	455.6875	449.1875
7	454.91875	448.41875	18	455.5375	449.0375
8	454.93125	448.43125	19	455.6250	449.1250
9	454.94375	448.44375			
10	454.95625	448.45625			
11	454.96875	448.46875			

Note; Channel 2 is the network signalling channel, i.e. Base TX only

National Bus Frequencies

These also use the Band III allocation and are interleaved with rail channels

Mobile TX	Mobile RX	Mobile TX	Mobile RX
201.31250	193.31250	201.81250	193.81250
201.36250	193.36250	201.86250	193.86250
201.41250	193.41250	201.91250	193.91250
201.51250	193.51250	201.96250	193.96250
201.56250	193.56250	202.01250	194.01250
201.61250	193.61250	202.06250	194.06250
201.66250	193.66250	202.11250	194.11250
201.71250	193.71250	202.16250	194.16250
201.76250	193.76250	202.21250	194.21250

Mobile TX	Mobile RX
202.26250	194.26250
202.31250	194.31250
202.36250	194.36250
202.46250	194.46250
202.51250	194.51250
202.96250	194.96250
203.01250	195.01250
203.06250	195.06250
203.11250	195.11250
203.16250	195.16250
203.21250	195.21250
203.26250	195.26250
203.31250	195.31250
203.36250	195.36250
203.66250	195.66250
203.71250	195.71250
203.76250	195.76250
203.81250	195.81250
203.86250	195.86250
203.91250	195.91250
204.71250	196.71250
204.76250	196.76250

Mobile TX	Mobile RX
204.81250	196.81250
204.86250	196.86250
204.91250	196.91250
204.96250	196.96250
205.06250	197.06250
205.46250	197.46250
205.51250	197.51250
205.56250	197.56250
205.61250	197.61250
205.61250	197.61250
205.66250	197.66250
205.71250	197.71250
205.76250	197.76250
205.86250	197.86250
205.91250	197.91250
206.01250	198.01250
206.06250	198.06250
206.11250	198.11250
206.16250	198.16250
206.21250	198.21250
206.26250	198.26250

Radio Mic Frequencies (all in MHz)
Licence Exempt channels;
173.8
174.1
174.5
174.8
175.0
News Gathering frequencies;
176.800
184.600
184.800
185.000
191.700
192.100
192.600
199.900
200.800
201.000
207.900
208.800
216.300
217.000
Shared Frequencies;
175.250
175.525
176.600
191.900
192.800
193.000
199.700
200.300
200.600
208.300
208.600
209.000
216.100
216.600
216.800
854.900
855.275
855.900

856.175
856.575
857.625
857.950
858.200
858.650
860.400
860.900
861.200
861.550
861.750
Co-ordinated frequencies;
176.400
177.000
192.300
200.100
207.700
208.100
855.475
856.500
855.075
855.675
857.025
857.775
856.375
857.500
858.375
856.800
858.250
860.725
857.425
858.700
861.375
860.600
859.100
861.925
859.525

119

Land mobile services

Under the banner of land mobile services is a varied range of communications users. In the following section you will find listed private mobile radio, emergency services, message handling and paging. The term 'land mobile' applies to any radio communications that takes place between either mobile-to-mobile or mobile-to-base, across land as opposed to air or marine. The mobile can either be a vehicle installation or a portable transceiver of the walkie-talkie type. Ranges of such equipment vary enormously. In open country, ranges of 20 or 30 miles are not unusual but in built-up areas this may be cut to considerably less. Users of mobile radio equipment operating in towns and cities often use antennas on very high buildings well away from the actual point of operation. Connection between the operator and the remote antenna site is usually through a dedicated landline or a radio/microwave link. Emergency services may have even more sophisticated arrangements with several antenna/transmitter sites to give total coverage of an area. This becomes particularly important when communication is to and from low powered handsets with limited antenna facilities.

Private mobile radio (PMR)

Private mobile radio is a form of communication between a base station and one or more mobile or portable units. Typical examples are the transceivers used by taxi firms. PMR is not to be confused with the government allocations and emergency services, all of which fall into different categories and are listed elsewhere. Communication in the PMR bands is primarily FM but some AM use remains, and may be split frequency, or single frequency simplex. Only one band, VHF band III is available solely for duplex working.

UK private mobile radio bands and frequency allocations

Frequency (MHz)	Allocation
55.7500-60.7500	UK low-band PMR
62.7500-68.0000	UK low-band PMR
68.08125-70.00625	Base / Simplex TX
71.50625-72.79375	Mobile TX
76.70625-77.99375	Mobile TX
81.50625-83.50000	Base TX
85.00625-87.49375	Base TX / Simplex
158.53125-160.54375	Base TX
163.03125-168.24375	Mobile TX
168.24375-168.30625	Simplex PMR
168.84375-169.39375	Simplex PMR
169.81875-173.09375	Mobile TX / Simplex PMR
177.20625-181.69375	Base TX PAMR/PMR
181.80625-183.49375	Base TX PAMR/PMR

Frequency (MHz)	Allocation
185.20625-189.69375	Mobile TX PAMR/PMR
189.80625-191.49375	Mobile TX PAMR/PMR
193.20625-199.49375	Mobile TX PAMR/PMR
201.20625-207.49375	Base TX PAMR/PMR
209.20625-210.20625	Base TX PAMR/PMR
212.55625-213.55625	Mobile TX PAMR/PMR
210.91875-211.91875	Base TX narrowband
214.26875-215.26875	Mobile TX narrowband
410.0000-415.0000	Mobile TX TETRA PMR
420.0000-425.0000	Base TX TETRA PMR
425.00625-427.75625	Mobile TX
428.01875-428.99375	Mobile TX
431.00000-432.00000	Mobile TX London only
440.00625-442.25625	Base TX
442.51875-443.49375	Base TX

Frequency (MHz)	Allocation
445.50625-446.40625	Base TX
447.51875-449.49375	Base TX London only
453.00625-453.99375	Mobile TX
454.8375-454.9875	Simplex
455.46875-455.85625	Airports only

Frequency (MHz)	Allocation
455.99375-456.99375	Mobile TX
459.51875-460.49375	Base TX
460.76875-461.23125	Airports only
461.25625-462.49375	Mobile TX and Simplex

PMR446

These are services designed for use with 500mW handhelds only for short-range use. PMR446 operates on 8 channels and is an EU harmonised allocation and may be used for any purpose. CTCSS or DCS is used on these services.
446.00625 MHz
446.01875 MHz
446.03125 MHz
446.04375 MHz
446.05625 MHz
446.06875 MHz
446.08125 MHz
446.09375 MHz

Short Term Hire Channels;

72.37500 MHz	85.87500 MHz (paired or simplex)	159.58750 MHz	164.08750 MHz (paired or simplex)
158.78750 MHz	163.28750 MHz (paired or simplex)	159.62500 MHz	164.12500 MHz (paired or simplex)
159.18750 MHz	163.68750 MHz (paired or simplex)	159.68750 MHz	164.18750 MHz (paired or simplex)
159.25000 MHz	163.75000 MHz (paired or simplex)	169.01250 MHz	(simplex)
		169.13750 MHz	(simplex)
159.35000 MHz	163.85000 MHz (paired or simplex)	169.16250 MHz	(simplex)
		169.18750 MHz	(simplex)
159.40000 MHz	163.90000 MHz (paired or simplex)	456.38750 MHz	461.88750 MHz (paired or simplex)
159.42500 MHz	163.92500 MHz (paired or simplex)	456.86250 MHz	462.36250 MHz (paired or simplex)
159.48750 MHz	163.98750 MHz (paired or simplex)	456.98750 MHz	462.48750 MHz (paired or simplex)
159.50000 MHz	164.00000 MHz (paired or simplex)	462.47500 MHz	(simplex)

UK 'General' PBR Frequencies
VHF Low Band
77.6875MHz
86.3375MHz
86.3500MHz
86.3625MHz
86.3750MHz
VHF Mid Band
164.0500MHz
164.0625MHz

VHF High Band
169.0875MHz
169.3125MHz
173.0875MHz
173.0625MHz
173.0500MHz
UHF
449.3125MHz
449.4000MHz
449.4750MHz

Amateur bands

Most general-purpose scanners will cover at least one of the VHF/UHF amateur bands. Although many scanner users may look to such things as air and marine bands as being the more exciting listening, amateur bands do have an attraction in that the operators are not subject to the same power restrictions, and so even at VHF and UHF amateur radio becomes international in its coverage. During the summer months, in particular, effects such as sporadic-E and tropospheric ducting can mean that signals can be picked up over several hundreds of miles. British amateurs use five bands on VHF and UHF; 6 metres, 4 metres, 2 metrse, 70 centimetres and 23 centimetres. There are other bands but these are beyond the coverage of most scanners.

6 metre band

Recommended UK frequency allocations in the 6 metre band

Frequency (MHz)	Allocation
50.000-50.100	CW only
50.020-50.080MHz -	beacons
50.100-50.500	SSB and CW only
50.110MHz	Intercontinental calling
50.500-51.000	All modes
50.710-50.910	FM Repeater outputs in 10kHz steps
51.000-52.000	All modes
51.210-51.410	FM repeater inputs (mobile TX)
51.430-51.830	FM simplex channels in 20kHz steps
51.510	FM simplex calling channel
51.830-52.00	All modes (emergency communications priority)

This particular band is also available to amateurs in countries in Regions 2 and 3 (including the USA). There, the band lies between 50- This is without a doubt the most popular amateur VHF band and signals can usually be heard on it in most areas at any time of day. The UK band extends from 144-146MHz but in other regions the band is extended even higher. Equipment for this band is relatively cheap and portable which makes it a favourite with amateurs for local contact work. Range on the band varies enormously. Varying conditions can mean that a transmission of several hundred watts output may only be heard 20 or 30 miles away at one time, while a signal of a few watts could be picked up hundreds of miles away at another time. Peak propagation tends to be in the summer when sporadic-E activity is at its highest. The band is used for a whole range of transmission types and several modes are used. Frequency allocation is more by a sort of gentlemen's agreement than anything else. The band is organised into blocks of transmission types. The abbreviation 'MS" stands for 'meteor scatter', a method of reflecting a radio signal off the tail of a meteor or a meteor shower. A similar method of communication is involved in 'moon-bounce'. These types of communications are generally beyond the scope of scanner users as highly sensitive equipment and massive antenna arrays are required. FM channelized simplex and repeater activity uses 12.5kHz spacing on this band although the majority of FM speech activity can be found on 25kHz spaced steps.

UK frequency allocations on the 2 metre band

Frequency (MHz)	Allocation
144.000-144.150	CW only
144.000-144.030MHz -	Moonbounce
144.050MHz -	CW calling
144.100MHz -	CW MS reference freq.
144.140-144.150 -	CW FAI working
144.150-144.500	SSB and CW only
144.150-144.160MHz -	SSB FAI working
144.195-144.205MHz -	SSB random MS
144.250MHz -	used for slow Morse transmissions and weekend news broadcasts
144.260MHz -	Emergency comms priority
144.300MHz -	SSB calling frequency
144.390-144.400MHz -	SSB random MS
144.500-144.800	All modes non-channelised
144.500MHz -	SSTV calling
144.575 -	ATV talkback (SSB)
144.700MHz -	FAX calling
144.750MHz -	ATV calling and talkback
144.775-144.800MHz -	Emergency communications priority
144.800-144.990	Digital modes
145.000-145.1875	FM repeater inputs
145.2000-145.5875	FM simplex channels
145.2000MHz -	V16 (S8) Emergency communications priority
145.2250MHz -	V18 (S9) Emergency communications priority
145.2500MHz -	V20 (S10) Used for slow Morse transmissions
145.2750MHz -	V22 (S11)
145.3000MHz -	V24 (S12) RTTY/AFSK
145.3250MHz -	V26 (S13)
145.3500MHz -	V28 (S14)
145.3750MHz -	V30 (S15)
145.4000MHz -	V32 (S16)
145.4250MHz -	V34 (S17)
145.4500MHz -	V36 (S18)
145.4750MHz -	V38 (S19)
145.5000MHz -	V40 (S20) FM Calling channel
145.5250MHz -	V42 (S21) Used for weekend news broadcasts
145.5500MHz -	V44 (S22) Used for rally and exhibition talk-in
145.5750MHz -	V46 (S23)
145.6000-145.7875MHz	FM repeater outputs
145.6000MHz -	RV48 (R0)
145.6125MHz -	RV49
145.6250MHz -	RV50 (R1)
145.6375MHz -	RV51
145.6500MHz -	RV52 (R2)
145.6625MHz -	RV53
145.6750MHz -	RV54 (R3)
145.6875MHz -	RV55
145.7000MHz -	RV56 (R4)
145.7125MHz -	RV57
145.7250MHz -	RV58 (R5)
145.7325MHz -	RV59
145.7500MHz -	RV60 (R6)
145.7625MHz -	RV61
145.7750MHz -	RV62 (R7)
145.7875MHz -	RV63
145.8000-146.0000	Satellite service

70 centimetre band

This band is allocated between 430.00 and 440MHz. It is allocated on a 'secondary' basis which means that amateurs using it must not interfere with other services on the band. Other users include PMR in the London area in the 431-432MHz segment, and military users across the range, who have primary status in this band. Short range low power telemetry devices, including some car key fobs, operate around 433.920MHz.

The characteristics of the band are very similar to those of the 2 metre band with the exception that operators do not get the extreme ranges achieved at times on the 2 metre band. By and large the band is less used than the 2 metre band although in densely populated areas there can be a fairly high level of activity. Like the 2 metre band, the 70 centimetre band also has repeaters, throughout the country, which considerably increase the range of operation.

Recommended UK frequency allocations in the 70 centimetre band

Frequency (MHz)	Allocation
430.000-432.000	All modes (431-432MHz is used for PMR in the London Area)
430.810-430.990	low power repeater inputs

Frequency (MHz)	Allocation
432.000-432.150	CW only
423.000-432.025MHz -	moonbounce
432.050MHz -	CW centre of activity
432.150-432.500	SSB and CW only
432.200MHz -	SSB centre of activity
432.350MHz -	Microwave talk-back
432.500-432.800	All modes non-channelised
432.500MHz -	SSTV activity centre
432.600MHz -	RTTY FSK activity centre
432.625, 432.650, 432.675MHz -	packet radio
432.700MHz -	FAX activity centre
432.800-432.990	Beacons
433.000-433.3875MHz	FM repeater outputs
433.000MHz -	RU240 (RB0)
433.025MHz -	RU242 (RB1)
433.050MHz -	RU244 (RB2)
433.075MHz -	RU246 (RB3)
433.100MHz -	RU248 (RB4)
433.125MHz -	RU250 (RB5)
433.150MHz -	RU252 (RB6)
433.175MHz -	RU254 (RB7)
433.200MHz -	RU256 (RB8)
433.225MHz -	RU258 (RB9)
433.250MHz -	RU260 (RB10)
433.275MHz -	RU262 (RB11)
433.300MHz -	RU264 (RB12)
433.325MHz -	RU266 (RB13)
433.350MHz -	RU268 (RB14)
433.375MHz -	RU270 (RB15)
433.400-434.600	FM simplex channels
433.400MHz -	U272 (SU160
433.425MHz -	U274 (SU17)
433.450MHz -	U276 (SU18)

Frequency (MHz)	Allocation
433.475MHz -	U278 (SU19)
433.500MHz -	U280 (SU20) FM calling channel
433.525MHz -	U282 (SU21)
433.550MHz -	U284 (SU22)
433.575MHz -	U286 (SU23)
433.600MHz -	U288 (SU24) RTTY AFSK
433.625-433.675MHz -	packet radio
433.700-433.775MHz -	Emergency communications priority
434.600-435.000	FM repeater inputs
435.000-438.000	Satellite service and FSTV
438.000-439.800	FSTV
438.900-440.00	Packet radio

23 centimetre band

A wide range of scanners, even the tiniest handheld types, now give coverage up to 1300MHz. This allows reception of the sections of the 23cm band used for FM simplex and repeater communication, as well as DX communication. The latter you will normally only hear during VHF/UHF contests, such as the annual VHF National Field Day. However 23cm repeaters are horizontally polarized, although this may change in the future. A number of amateurs use 23cm to get away from the relatively congested 2m and 70cm bands, although activity is currently very low due to the high cost of equipment. However, as this comes down in price, occupancy is likely to increase.

Recommended UK frequency allocations in the 23 centimetre band

Frequency (MHz)	Allocation
1240.000-1243.250	All modes
1240.150-1240.750MHz	digital communications
1243.250-1260.000	ATV
1248.000MHz	RMT1-3 TV repeater input
1249.000MHz	RMT1-2 TV repeater input
1260.000-1270.000	Satellite service
1270.000-1272.000	All modes
1272.000-1291.500	ATV
1276.500	RMT1-1 AM TV repeater input
1291.000-1291.500	Repeater inputs
1291.500-1296.000	All modes
1296.000-1296.150	CW
1296.150-1296.800	SSB
1296.200MHz	centre of narrowband activity
1296.500MHz -	SSTV
1296.600MHz -	RTTY
1296.700MHz -	Fax
1296.800-1297.990	Beacons
1297.000-1297.475	Repeater outputs 25kHz spacing
1296.000MHz -	RM0
1296.025MHz -	RM1
1296.050MHz -	RM2
1296.075MHz -	RM3
1296.100MHz -	RM4
1296.125MHz -	RM5
1296.150MHz -	RM6
1296.175MHz -	RM7
1296.200MHz -	RM8
1296.225MHz -	RM9
1296.250MHz -	RM10

Frequency (MHz)	Allocation
1296.275MHz -	RM11
1296.300MHz -	RM12
1296.325MHz -	RM13
1296.350MHz -	RM14
1296.375MHz -	RM15
1297.500-1298.000	FM Simplex
1297.500MHz -	SM20
1297.525MHz -	SM21
1297.550MHz -	SM22
1297.575MHz -	SM23
1297.600MHz -	SM24
1297.625MHz -	SM25
1297.650MHz -	SM26
1297.675MHz -	SM27
1297.700MHz -	SM28
1297.725MHz -	SM29
1297.750MHz -	SM30
1298.000-1298.500	All modes (digital communications)
1298.500-1300.000	Packet radio
1299.000MHz	Packet radio 25kHz bandwidth
1299.425MHz	Packet radio 150kHz bandwidth
1299.575MHz	Packet radio 150kHz bandwidth
1299.725MHz	Packet radio 150kHz
1300.000 1325.000	TV repeater outputs
1308.000MHz	RMT1-3 FM TV repeater output
1311.500MHz	RMT1-1 AM TV repeater output
1316.000MHz	RMT1-2 FM TV repeater output

Wide area paging

This service provides for one way transmissions from a base station to a small pocket receiver. The transmission is coded to activate only the required pager, and the most common system used is POCSAG, although the FLEX proprietary protocol is also used, ERMES is a pan-European paging system. The simplest form of pagers merely emit a bleeping sound to alert the holder that they are wanted. Some of the more sophisticated types can receive a short alphanumeric message that appears on a small liquid crystal display. Wide area pagers usually cover a specific area such as a town but a number of services also cover most of the country. Wide area paging should not be confused with 'on-site' paging.

Wide Area Paging Allocated frequencies
VHF
137.9625-138.2125MHz
153.0125-153.4875MHz
169.4125-169.8125MHz (ERMES)
UHF
454.0125-454.8375MHz
466.0625-466.0875MHz

On-site paging

Similar to wide area paging but low powered and operating over a small area such as a factory, building site, etc. Sometimes the pager has a small and simple transmitter which allows the user to acknowledge that the paging signal has been received. AM or FM modes may be transmitted, and data communications are possible. At VHF a 12.5kHz channel spacing is used, at UHF 25kHz.

On-site paging allocated frequencies

VHF

31.7125-31.7875MHz	Hospital paging
47.400MHz	Vehicle paging alarms
48.9750-49.49375	On-site paging
49.4250MHz	Hospitals
49.4375MHz	Hospitals
49.4500MHz	Hospitals
49.4625MHz	Hospitals
49.4750MHz	Hospitals
160.99375-161.20625MHz	On-site, with return speech acknowledgment allowed for hospital in emergencies

UHF

458.8375-459.4875MHz	On-site
458.900MHz	Vehicle paging and some car radio keyfobs

Land emergency services

Land emergency services are increasingly migrating over to TETRA, virtually all police forces use this as primary communication and Fire and Rescue services have been migrating to this also for base to mobile communication, although 'fire ground' communication, i.e. between personnel at the incident, is normal analogue FM (frequencies are listed below), often cross-linked to TETRA for intercommunication with other emergency services involved at the incident.

In addition to the bands listed below, emergency services in some areas may be located in government mobile allocations.

Land emergency services' band and allocated frequencies

Band: VHF low (12.5kHz channel spacing, AM)

Frequency (MHz)	Allocation
70.5000-71.5000	Fire bases
81.9000 83.9000	Fire mobiles

Band; VHF High (12.5kHz channel spacing, FM)

Frequency (MHz)	Allocation
166.2750-166.5250	Ambulances

Ambulance services

Area	Base	Mobile	Channel
Avon	166.5000	171.3000	117
Bedfordshire	166.3375	171.1375	104
Bedfordshire	166.4625	171.2625	114
Bedfordshire	166.7750	171.5750	139
Berkshire	166.3875	171.1875	108
Berkshire	166.6125	171.4125	126
Buckinghamshire	166.2875	171.0870	100
Buckinghamshire	166.5625	171.3625	122
Cambridgeshire	166.3125	171.1125	102

Area	Base	Mobile	Channel
Cambridgeshire	166.3500	171.1500	105
Cheshire	166.3625	171.1625	106
Cleveland	166.2000	171.0000	93
Cleveland	166.3500	171.1500	105
Clwyd	166.4125	171.2125	110
Clwyd	166.4625	171.2625	114
Clwyd	166.5625	171.3625	122
Cornwall	166.2875	171.0870	100
Cornwall	166.5000	171.3000	105
Cumbria	166.3750	171.1750	107
Derby	166.2875	171.0870	100
Derby	166.3125	171.1125	102
Derbyshire	166.3750	171.1750	107
Devon	166.3125	171.1125	102
Devon	166.5625	171.3625	122
Doctors common	166.8125	171.6125	1
Dorset	166.2000	171.0000	93
Dorset	166.3000	171.1000	101
Dorset	166.4875	171.2875	116
Dorset	166.5250	171.3250	119
Dorset	166.8375	171.6375	144
East Anglia	166.0500	170.8500	
East Anglia	166.3500	171.1500	105
East Anglia	166.3625	171.1625	106
East Anglia	166.4375	171.2375	112
East Anglia	166.5250	171.3250	119
Essex	166.3625	171.1625	106
Essex	166.4875	171.2875	116
Essex	166.5500	171.3500	121
Glamorgan	166.7750	171.5750	139
Glamorgan (mid)	166.3250	171.1250	103
Glamorgan (mid)	166.5250	171.3250	119
Glamorgan (S)	166.3000	171.1000	101
Glamorgan (S)	166.4750	171.2750	115
Glamorgan (S)	166.5875	171.3875	124
Glamorgan (W)	166.3500	171.1500	105
Glamorgan (W)	166.8250	171.6250	143
Gloucestershire	166.3625	171.1625	106
Gloucestershire	166.8000	171.6000	142
Gloucestershire	166.8375	171.6375	144
Guernsey	86.4250		
Gwent	166.4000	171.2000	109
Gwent	166.5750	171.3750	123
Gwynedd	166.4750	171.2750	115
Hampshire	166.3625	171.1625	106
Hampshire	166.5750	171.3750	123
Hampshire	166.5875	171.3875	124
Hampshire	166.7750	171.5750	139
Hatfield	166.6125	171.4125	2
Hatfield	166.8125	171.6125	1
Herefordshire	166.3750	171.1750	107
Herefordshire	166.4750	171.2750	115
Herefordshire	166.5625	171.3625	122
Herefordshire	166.6125	171.4125	126
Herefordshire	166.8250	171.6250	143
Hertfordshire	166.5875	171.3875	124
Humberside	166.3000	171.1000	101
Humberside	166.3250	171.1250	103
Humberside	166.4000	171.2000	109
Humberside	166.5250	171.3250	119
Humberside	166.5750	171.3750	123
Humberside	166.6125	171.4125	126
Isle of Wight	166.3375	171.1375	104
Isles of Scilly	166.5000	171.3000	117
Jersey	154.6625	146.1125	1

Area	Base	Mobile	Channel
Jersey	154.7500	146.2250	2
Kent	166.2875	171.0875	100
Kent	166.3375	171.1375	104
Kent	166.3875	171.1875	108
Kent	166.8250	171.6250	143
Lancashire	166.2750	171.0750	99
Lancashire	166.3875	171.1875	108
Lancashire	166.5500	171.3500	121
Leicestershire	166.4125	171.2125	110
Leicestershire	166.5375	171.3375	120
Leicestershire	166.3000	171.1000	101
Lincolnshire	166.2000	171.0000	93
Lincolnshire	166.2750	171.0750	99
Lincolnshire	166.3625	171.1625	106
Lincolnshire	166.4000	171.2000	109
London	165.6250	170.4250	3
London	165.6375	170.4750	2
London	165.6500	170.8500	1
London	166.1000	170.9000	6
London	166.1250	170.9250	5
London	166.3125	171.1125	1
London	166.4375	171.2375	15
London	166.5250	171.3250	4
London	166.5750	171.3750	10
London (E)	166.3500	171.1500	2
London (NE)	166.3000	171.1000	9
London (NE)	166.4250	171.2250	7
London (NE)	166.5000	171.3000	8
London (NW)	166.4500	171.2500	6
London (NW)	166.4750	171.2750	5
London (S)	166.4125	171.2125	3
London (SE)	166.2000	171.0000	13
London (SE)	166.3250	171.1250	12
London (SE)	166.3750	171.1750	14
London (SW)	166.2750	171.0750	11
Manchester	166.2875	171.0870	100
Manchester	166.3000	171.1000	101
Manchester	166.4875	171.2875	116
Manchester	166.5000	171.3000	117
Manchester	166.5125	171.3125	118
Manchester	166.6000	171.4000	125
Merseyside	166.3250	171.1250	103
Merseyside	166.3375	171.1375	104
Merseyside	166.4750	171.2750	115
Merseyside	166.5875	171.3875	124
Midlands (W)	166.2750	171.0750	99
Midlands (W)	166.3500	171.1500	105
Midlands (W)	166.4625	171.2625	114
Midlands (W)	166.5500	171.3500	121
Midlands (W)	166.6000	171.4000	125
National	87.6500	77.6500	
National	166.1000	170.9000	
National	166.6000	171.4000	
National	166.6125	171.4125	
Norfolk	166.4375	171.2375	1
Norfolk	166.5625	171.3625	2
Northern Ireland	87.5500	77.5500	
Northern Ireland	87.5750	77.5750	
Northern Ireland	87.6250	77.6250	
Northern Ireland	87.6500	77.6500	
Northern Ireland	87.6750	77.6750	
Northern Ireland	87.5250	77.5250	
Northumbria	166.2875	171.0870	100
Northumbria	166.4000	171.2000	109
Northumbria	166.4875	171.2875	116

Area	Base	Mobile	Channel
Northumbria	166.5125	171.3125	118
Northumbria	166.5750	171.3750	123
Northumbria	166.6000	171.4000	125
Northamptonshire	166.5500	171.3500	121
Northamptonshire	166.5750	171.3750	123
Nottinghamshire	166.4125	171.2125	110
Oxfordshire	166.4875	171.2875	116
Oxfordshire	166.6125	171.4125	126
Pembrokeshire	166.3625	171.1625	106
Powys	166.2875	171.0870	100
Powys	166.3125	171.1125	102
Private national	72.5375		
Private national	86.0375		
Red Cross	86.3250		
Red Cross	86.4125		
Somerset	166.2750	171.0750	99
Somerset	166.3375	171.1375	104
Staffordshire	166.3875	171.1875	108
Staffordshire	166.5000	171.3000	117
Staffordshire	166.5875	171.3875	124
Staffordshire	166.6125	171.4125	126
Staffordshire	166.7500	171.7500	137
Suffolk	166.5250	171.3250	119
Suffolk	166.3500	171.1500	105
Suffolk (W)	166.3375	171.1375	104
Surrey	166.2875	171.0870	100
Surrey	166.5125	171.3125	118
Surrey	166.5375	171.3375	120
Surrey	166.7500	171.7500	137
Sussex (E)	166.4000	171.2000	109
Sussex (W)	166.4625	171.2625	114
Sussex (W)	166.5625	171.3625	122
Welwyn	166.6125	171.4125	2
Welwyn	166.8125	171.6125	1
Wiltshire	166.4125	171.2125	110
Wiltshire	166.6125	171.4125	126
Worcestershire	166.3750	171.1750	107
Worcestershire	166.4750	171.2750	115
Worcestershire	166.5625	171.3625	122
Worcestershire	166.8250	171.6250	143
Yorkshire	166.2000	171.0000	93
Yorkshire	166.3875	171.1875	108
Yorkshire (N)	166.3375	171.1375	104
Yorkshire (N)	166.4625	171.2625	114
Yorkshire (N)	166.4750	171.2750	115
Yorkshire (N)	166.5000	171.3000	117
Yorkshire (N)	166.5250	171.3250	119
Yorkshire (N)	166.5375	171.3375	120
Yorkshire (N)	166.6125	171.4125	126
Yorkshire (S)	166.4000	171.2000	109
Yorkshire (S)	166.4875	171.2875	116
Yorkshire (S)	166.5500	171.3500	121
Yorkshire (S)	166.5625	171.3625	122
Yorkshire (W)	166.3000	171.1000	101
Yorkshire (W)	166.4000	171.2000	109
Yorkshire (W)	166.4125	171.2125	110
Yorkshire (W)	166.4625	171.2625	114
Yorkshire (W)	166.5375	171.3375	120

UHF allocations for Regional Health Authorities

Frequency	Channel	Area	Frequency	Channel	Area
457.4000	2	Yorkshire	461.350	10	Oxford Wales
457.4250	2	Yorkshire	461.375	10	Oxford
457.4250	4	East Anglia	461.475	10	Oxford
457.4500	1	Northern			Wales
457.4750	4	East Anglia	462.925	4	East Anglia
457.5000	4	East Anglia	462.950	1	Northern
457.5250	12	West Midlands		9	Wessex
457.5750	12	West Midlands			Wales
457.6250		Wales	462.975	4	East Anglia
457.6750			463.000		
166.335	4	East Anglia	166.525	4	East Anglia
457.7250	1	Northern	463.050	10	Oxford
457.7500		Wales	463.075	4	East Anglia
457.7750	4	East Anglia	463.100	1	Northern
457.9750	4	East Anglia			Wales
458.0000	2	Yorkshire	463.150	1	Northern Wales
458.0250	5,6,7,8	Thames	463.175		
458.0500	3	Trent	166.3375	4	East Anglia
	11	South West	463.225		Wales
458.1250	1	Northern	463.250	4	East Anglia
458.1500	12	West Midlands	463.500	9	Wessex
458.1750	2	Yorkshire	463.525	1	Northern
	12	West Midlands		4	East Anglia
		Wales	463.550		Wales
458.2250	4	East Anglia	463.600	9	Wessex
		Wales			Wales
458.2500	2	Yorkshire	463.625	1	Northern
458.3000		Wales		9	Wessex
458.3250	1	Northern		12	West Midlands
	5 6 7 8	Thames	463.650	5 6 7 8	Thames
458.3500	1	Northern	463.700	12	West Midlands
458.4250	5 6 7 8	Thames	463.750	9	Wessex
458.4750		Wales		12	West Midlands
459.7750		Jersey	463.825	1	Northern Wales
460.5750	5 6 7 8	Thames	463.850	1	Northern Wales
	11	South West	463.900	5 6 7 8	Thames
460.6000	10	Oxford	463.925	1	Northern Wales
460.6250	1	Northern	463.950	5 6 7 8	Thames
	10	Oxford		9	Wessex
460.6500		Wales		12	West Midlands
460.675	1	Northern	463.975	12	West Midlands
	2	Yorkshire	467.025	12	West Midlands
460.725	11	South West	467.050		Wales
460.750	1	Northern	467.125	1	Northern
461.275	1	Northern	467.175	1	Northern
	2	Yorkshire	467.225	11	South West
	5 6 7 8	Thames	467.250	1	Northern Wales
	12	West Midlands	467.475		Wales
		Wales	467.775		Wales
461.300	2	Yorkshire	467.800		Wales
	5 6 7 8	Thames	467.825		Wales
	10	Oxford Wales	467.900		Wales
461.325	10	Oxford	467.925		Wales

Areas by number

1 Northern: Cleveland, Cumbria, Durham, Northumbria
2 Yorkshire: Yorkshire, Humberside
3 Trent: Derbyshire, Leicestershire, Lincolnshire, Nottinghamshire, S. Yorkshire
4 East Anglia: Cambridgeshire, Norfolk, E. Suffolk
5 NW Thames: N. Bedfordshire, E. Hertfordshire
6 NE Thames: Mid Essex
7 SE Thames: Eastbourne, Medway, Kent
8 SW Thames: Mid Surrey, W. Sussex
9 Wessex: E. Dorset, Hampshire, Isle of Wight
10 Oxford: W. Berkshire, Buckinghamshire, Northamptonshire, Oxford
11 South West: Avon, Cornwall, Devon, Gloucester, Scilly Isles, Somerset
12 West Midlands: Hereford, Mid Staffordshire, Salop, South Warwickshire, Worcestershire
13 Mersey: Cheshire, Merseyside

SCANNERS

Fireground UHF Channels

These are use 'at scene' at all locations in the UK during fire and rescue incidents

457.0125   FM   Fireground Chan 3
457.0375   FM   Fireground Chan 1
457.0875
462.5875   FM   Fireground Chan 2
457.1375
462.6375   FM   Fireground Chan 5
457.1875   FM   Fireground Chan 4
457.2375   FM   Fireground Chan 6
457.4875   FM   Fireground Chan 7
418.8375   FM   MOD National (used during firefighting support)

Fire Brigades, VHF (mobile base to tender - note; many counties are gradually going over to TETRA)

Band: VHF High (12.5kHz channel spacing (AM/FM)

Band Allocation

70.5125-71.5000 Fire bases
80.0000-82.5000 Fire mobiles
146.0000-148.000 Fire mobiles
154.0000-156.000 Fire bases

Area	Base	Mobile	Call	Channel
Avon & Somerset	71.0125	80.1750	M2QG	
Avon & Somerset	154.5000		M2QC	
Bedfordshire	71.1125		M2VM	
Berkshire	154.0750	146.1750	M2HD	
Berkshire	71.2000		M2HD	
Buckinghamshire	71.1375	80.1125	M2HK	
Cambridgeshire	154.1250	146.1750	M2VP	
Cambridgeshire	70.8375		M2VC	
Cambridgeshire	71.4250		M2VC	
Cheshire	154.2250		M2CF	
Cheshire	154.6500		M2CF	
Cheshire	155.5750	146.2875	M2CF	
Cheshire	70.7750	80.5000	M2CF	
Cleveland	71.1125		M2LT	
Clwyd	71.1625	80.8750	M2WK	
Cornwall	70.7875	80.8000	M2QA	
County Durham	70.8875	80.2125	M2LF	
Cumbrin	70.8375	80.0375	M2BC	
Derbyshire	154.0500	146.5375	M2ND	
Derbyshire	70.7125	80.8000	M2ND	
Devon	70.7250	80.0375	M2QD	
Devon	70.8250	80.0375	M2QD	
Devon & Cornwall	154.0750	146.1750	M2QA	
Dorset	70.8625	80.5500	M2QK	
Dyfed	70.6125	80.1250	M2WV	
East Anglia	154.6500	146.1125		
East London	70.7625	80.1500	M2FE	3
Essex	70.6250	80.6125	M2VD	1
Essex	70.7250	80.6750	M2VD	2
Essex	70.9125		M2VD	
Gloucestershire	71.0750	80.6250	M2QF	
Gloucestershire	71.3875		M2QC	
Gloucestershire	154.3125		M2QF	2
Greater Manchester	70.5875	80.7625	M2FT	3
Greater Manchester	70.5250	80.7375	M2FT	2
Greater Manchester	70.5500		M2FT	
Greater Manchester	70.8250	80.7875	M2FT	4
Gwent	70.7000	81.1250	M2WP	
Gwynedd	70.8125	81.2125	M2WC	
Hampshire	154.5750		M2HX	
Hampshire	154.7625		M2HX	
Hampshire	155.3375		M2HX	
Hampshire	70.5875	80.1875	M2HX	2
Hampshire	70.7750	80.5000	M2HX	1

132

Area	Base	Mobile	Call	Channel
Hampshire (N)	154.8750		M2ND	
Handsets	154.6875			
Hereford & Worcester	70.6875	81.1250	M2YB	
Hereford & Worcester	154.3000		M2YB	
Hereford & Worcester	155.4000		M2YB	
Hertfordshire	70.9000	80.0375	M2VI	
Humberside	154.6000		M2XT	
Humberside	154.8250	146.7000	M2XT	
Humberside	71.0750		M2XT	
Humberside	71.1000	80.6625	M2XT	
Humberside (N)	154.7750		M2XT	
Isle of Wight	71.2750	81.0625	M2HP	
Kent	154.4750	146.1750	M2HO	5
Kent	154.7500		M2KA	1
Kent	70.8375	80.0375	M2KF	
Lancashire	70.5250	80.9625	M2MP	2
Lancashire	71.3875		M2BE	3
Lancashire	70.6750	80.5500	M2BE	1
Lancashire	70.9000	80.6000	M2BE	4
Leamington	155.2250		M2YS	
Leicestershire	70.6625	81.0000	M2NK	
Lincolnshire	70.5625 8	0.6000	M2NV	
London	154.1250	146.1750	M2FN	
London	154.1750	146.3500	M2FS	2
London	154.6750	146.2750	M2FE	3
London	154.8250	147.6125	M2FN	4
London	71.3000		M2FH	5
London (N)	71.1750	80.2125	M2FN	4
London (S)	70.9626	80.1125	M2FS	2
London Central	70.5250	80.9625	M2FH	1
Merseyside	70.6250		M2F0	3
Merseyside	70.7000		M2F0	4
Merseyside	70.9625	81.0875	M2FO	2
Merseyside	71.0375	81.0625	M2FO	
Mid Glamorgan	70.5625	80.6000	M2WF	
Norfolk	70.7000		M2VF	
Norfolk	154.4000		M2VF	
Norfolk	154.7250		M2VF	
Norfolk	154.8500		M2VF	
North Yorkshire	71.1375	80.4375	M2LY	
North Yorkshire	71.1750		M2LY	
Northamptonshire	70.7500	80.7500	M2N0	
Northumberland	70.5125	80.1875	M2LJ	
Nottinghamshire	154.6750	146.2750	M2NZ	
Nottinghamshire	70.5375	80.1875	M2NZ	
Oxfordshire	71.1000	80.6625	M2HI	
Powys	70.8500	80.9625	M2WB	
Shropshire	70.9750	80.6500	M2YU	
Somerset	71.1250	80.1125	M2QI	
South Glamorgan	70.6750	80.5250	M2WD	
South Yorkshire	70.6375	80.2125	M2XY	
Staffordshire	154.1250	146.1750	M2YG	
Staffordshire	154.2250	146.7000	M2YG	
Staffordshire	154.4750	146.1750	M2YG	
Staffordshire	155.6000		M2YG	
Staffordshire	70.8875	80.9375	M2YG	
Suffolk	71.2750	81.0875	M2VN	
Surrey	70.6125	80.1250	M2HF	
Surrey	146.0250		M2HF	
Sussex	70.6375	80.2125	M2KD	
Sussex (E)	154.8250	146.6125	M2KW	
Sussex (W)	70.8000	80.5125	M2KW	
Sussex (W)	154.8000	146.7250	M2KW	
Thames Valley	154.0750	146.1750	M2HI	
Thames Valley	154.5500		M2HI	

Area	Base	Mobile	Call	Channel
Tyne and Wear	71.3000		M2LP	
Tyne and Wear	71.3000		M2LP	2
Wales 154.8750	146.0375		M2WP	
Wales	154.9500	146.0125	M2WP	
Warwickshire	70.6000	80.2625	M2YS	
West Glamorgan	70.9500		M2WZ	
West Midlands	70.5125	80.4375	M2BW	1
West Midlands	154.2500		M2FB	
West Midlands	155.4500		M2FB	
West Midlands	70.5750	80.4625	M2F0	3
West Midlands	71.1500	80.5125	M2EW	2
West Yorkshire	70.7625		M2XF	
West Yorkshire	70.6125		M2XF	
West Yorkshire	70.8750		M2XF	
Wiltshire	154.1250	146.1750	M2QM	
Wiltshire	154.4750	146.1750	M2QM	
Wiltshire	70.6500	80.9875	M2QM	
Yorkshire	154.0500	146.5375	M2XK	

Citizens' Band (CB) radio

Across open country usable HF ranges of up to 20 or more miles can be expected but a lot will depend on circumstances, i.e., base/base, base/mobile or mobile/mobile working. Ranges are considerably reduced in built-up areas but under lift conditions ranges may become almost global. The band is very prone to the effects of the 11 year sunspot cycle and when that happens it is not unusual to hear transmissions from the USA, Australia, Asia, South America, etc. On the UHF band, activity is usually limited to local conditions, but even at these high frequencies lift conditions do occur. For instance, cross channel contacts between stations in southern England and the Channel Islands (distances of more than 100 miles) regularly take place in the summer months. Note that ranges quoted are only likely to be achieved by using a proper CB antenna. Most discone-type antennas, favoured for scanner operation, operate quite poorly at lower frequencies.

Pan-European CB channels (FM)

These are the CEPT channels which are common to Europe, including the UK, although UK-specific channels are also in use (see below)

Chan	Freq (MHz)	Chan	Freq (MHz)
01	26.965	15	27.135
02	26.975	16	27.155
03	26.985	17	27.165
04	27.005	18	27.175
05	27.015	19	27.185
06	27.025	20	27.205
07	27.035	21	27.215
08	27.055	22	27.225
09	27.065	23	27.255
10	27.075	24	27.235
11	27.085	25	27.245
12	27.105	26	27.265
13	27.115	27	27.275
14	27.125	28	27.285

Chan	Freq (MHz)		Chan	Freq (MHz)
29	27.295		35	27.355
30	27.305		36	27.365
31	27.315		37	27.375
32	27.325		38	27.385
33	27.335		39	27.395
34	27.345		40	27.405

UK -Specific CB Radio Channels

These are UK-specific channels which will officially remain available until July 2010, however due to the vast numbers of radios already in use it is envisaged these will still be in use by many for several years to come.

Chan	Freq (MHz)		Chan	Freq (MHz)
01	27.60125		21	27.80125
02	27.61125		22	27.81125
03	27.62125		23	27.82125
04	27.63125		24	27.83125
05	27.64125		25	27.84125
06	27.65125		26	27.85125
07	27.66125		27	27.86125
08	27.67125		28	27.87125
09	27.68125		29	27.88125
10	27.69125		30	27.89125
11	27.70125		31	27.90125
12	27.71125		32	27.91125
13	27.72125		33	27.92125
14	27.73125		34	27.93125
15	27.74125		35	27.94125
16	27.75125		36	27.95125
17	27.76125		37	27.96125
18	27.77125		38	27.97125
19	27.78125		39	27.98125
20	27.79125		40	27.99125

174MHz Wireless Microphone Frequencies

Colour	Frequency
Yellow	173.800MHz
Red	174.100MHz
Blue	174.5MHz
Green	174.800MHz
White	175.000MHz

National and Local Paging channels

Freq (MHz)	Use
147.800	Fire Brigade Alerter Paging (Main channel)
147.825	Fire Brigade Alerter Paging
147.850	Fire Brigade Alerter Paging
138.5125 - 139.5000	Mercury, Vodapage (12.5kHz steps)
153.025	Chan 1 National paging
153.050	Chan 2 National paging (Fire Brigades)
153.075	Chan 3 National paging
153.100	Chan 4 National paging
153.125	Chan 5 Redifon national paging
153.150	Chan 6 Redifon national paging
153.175	Chan 7 BT national paging

Freq (MHz)	Use
153.200	Chan 8 National paging
153.225	Chan 9 Redifon national paging
153.250	Chan 10 National paging
153.275	Chan 11 National paging
153.300	Chan 12 National paging
153.325	Chan 13 National paging
153.350	Chan 14 Mercury national paging
153.375	Chan 15 National paging
153.400	Chan 16 UKAEA (UK Atomic Energy Authority) national paging
153.425	Chan 17 National paging
153.450	Chan 18 National paging
153.475	Chan 19 National paging
153.500	Chan 20 National paging

135

Freq (MHz) Use

161.0125 - 161.1125 On-site voice paging
(12.5kHz steps)
453.8875   Benefits Agency paging channel
454.025 - 454.050   National Voice paging
454.125   Thames Coastguard to RNLI
paging
454.825   Page Boy national paging
455.300   Page Boy national paging
459.150 - 149.475   On-site voice paging
466.075   Hutchison Telecom national
paging

National Road Construction Channels

RX	TX	Use
82.5250 MHz	69.0250 MHz	Traffic Management
163.0875 MHz	158.4875 MHz	Construction
163.3000 MHz	158.8000 MHz	Construction
163.3750 MHz	158.8750 MHz	Construction
163.5125 MHz	159.0125 MHz	Construction
163.6125 MHz	159.1125 MHz	Construction
165.0750 MHz	169.8750 MHz	Engineering

Shopwatch Channels

Base TX	Base RX	Location
453.0250	459.52500	Aberavon
456.7500	462.2500	Aberdare
456.6500	462.1500	Aberdeen
456.6750	462.1750	Abingdon
453.0625	459.5625	Aldershot
443.0000	428.5000	Altrincham
456.6750	462.1750	Ashford
456.3500	461.8500	Ashton Under Lyne
456.5250	462.0250	Aylesbury
456.3500	461.8500	Ayr
441.0375	426.5375	Barking
453.9250	460.4250	Barnsley
456.8250	462.3250	Barnstaple
456.6500	462.1500	Barrow in Furness
456.5250	462.0250	Barry (S Wales)
453.4500	459.9500	Basildon
453.8000	460.3000	Bath Avon
456.3500	461.8500	Bedford
453.4750	459.9750	Bedworth
453.7750	460.2750	Belfast
456.9500	462.4500	Bexleyheath
456.6750	462.1750	Billericy
453.6000	460.1000	Bilston
453.7750	460.2750	Birkenhead
456.9500	462.4500	Birmingham Acocks Green
441.6500	427.1500	Birmingham Bearwood
453.7000	460.2000	Birmingham Kings Heath
453.4750	459.9750	Bispham and Cleveleys
456.6500	462.1500	Blackburn
453.1000	459.6000	Blackpool
445.7000	425.2000	Blantyre

Base TX	Base RX	Location
453.8500	460.3500	Blyth
453.9750	460.4750	Bognor Regis
440.9500	426.4500	Bolton
441.0375	426.5375	Bootle
453.0625	459.5625	Boscombe
456.5500	462.0500	Boston
453.0875	459.5875	Bournemouth
456.6750	462.1750	Braintree
453.0250	459.5250	Brentford
453.7500	460.2500	Bridgend Derwen
456.0250	461.5250	Bridlington
453.0500	459.5500	Bridport
456.5750	462.0750	Brighton
456.0500	461.5500	Brighton London Road
456.5250	462.0250	Bristol
166.7625	171.5625	Burnley
453.4750	459.9750	Burton on Trent
453.1000	459.6000	Bury St Edmunds
453.1000	459.6000	Caerphily
453.1750	459.6750	Camarthen
456.6000	462.1000	Cannock
166.7625	171.5625	Canterbury
453.2250	459.7250	Cardiff
456.3500	461.8500	Cardiff
456.6000	462.1000	Cardiff
453.7000	460.2000	Cardiff Canton
453.9875	460.4875	Carlisle
453.0625	459.5625	Chatham
456.8250	462.3250	Cheltenham
165.9250	170.7250	Chester le Street
453.0625	459.5625	Chesterfield
453.6000	460.1000	Chipping Norton
453.4750	459.9750	Chorley
453.0625	459.5625	Chorlton cum Hardy

Base TX	Base RX	Location
453.0875	459.5875	Cirencester
456.0250	461.5250	Clacton on Sea
456.0250	461.5250	Cleethorpes
443.0625	428.5625	Coalville
456.8500	462.3500	Colchester
453.7500	460.2500	Coleraine
453.7000	460.2000	Cosham
453.0875	459.5875	Crawley
453.0875	459.5875	Crewe
166.7625	171.5625	Crosby
453.0250	459.5250	Croydon
456.0250	461.5250	Cumbernauld
456.9750	462.4750	Dagenham
453.8000	460.3000	Dalkieth
456.6000	462.1000	Darlington
443.0500	428.5500	Dartford
456.5250	462.0250	Denton
456.7500	462.2500	Derby
456.6000	462.1000	Dover
453.5000	460.0000	Droylesden
456.9500	462.4500	Dudley
453.7750	460.2750	Dumfries
456.0250	461.5250	Durham
453.9750	460.4750	Durham Bishop Auckland
456.3500	461.8500	Ealing
456.5250	462.0250	Eastbourne
456.6750	462.1750	Ebbw Vale
441.0375	426.5375	Edgware
446.2625	446.2625	Edinburgh Craigleith
440.9125	426.4125	Edinburgh Kinnaird Park
453.5250	453.5250	Edinburgh Kirkgate
456.7500	462.2500	Edinburgh St James
440.1750	425.6750	Ellesmere Port
453.9250	460.4250	Fareham
453.9250	460.4250	Fareham
456.7500	462.2500	Farnborough
456.9500	462.4500	Farnham
442.6500	428.1500	Farnworth
453.2500	459.7500	Felixstowe
453.7000	460.2000	Fleetwood
456.6500	462.1500	Folkestone
456.5250	462.0250	Gillingham
166.7625	171.5625	Gloucester
456.5500	462.0500	Gloucester
453.0250	459.5250	Gorlestone
456.7500	462.2500	Grantham
167.0375	171.8375	Gravesend
456.6000	462.1000	Greenock
453.0875	459.5875	Guildford
453.0875	459.5875	Halesowen
453.6000	460.1000	Halifax
166.8750	171.6750	Hampstead London
456.6750	462.1750	Harlow
166.7625	171.5625	Harrogate
453.1125	459.6125	Harrow
453.7750	460.2750	Hartlepool
456.6750	462.1750	Harwich
453.7000	460.2000	Hastings
453.1000	459.6000	Havant
453.4750	459.9750	Havant
456.8750	462.3750	Havant Leigh Park
453.1750	459.6750	Hemel Hempsted

Base TX	Base RX	Location
453.0625	459.5625	Hereford
453.7000	460.2000	High Wycombe
456.6500	462.1500	Hitchin
456.3500	461.8500	Holyhead
453.4750	459.9750	Hounslow
456.8500	462.3500	Hove
456.9500	462.4500	Huddersfield
453.8250	453.8250	Hull
453.4750	459.9750	Huntingdon
456.9500	462.4500	Ilford
453.0500	459.5500	Ilkeston
166.7625	171.5625	Irvine Rivergate
453.0875	459.5875	Irvine
456.9500	462.4500	Kendal
453.8000	460.3000	Kidderminster
453.4500	459.9500	Kilmarnock
453.0500	459.5500	Kings Lynn
445.8750	425.3750	Kingston Upon Thames
166.7625	171.5625	Kirkcaldy Mercat
456.9500	462.4500	Kirkintilloch
453.7000	460.2000	Leamington Spa
456.7250	462.2250	Leeds
453.6000	460.1000	Leeds Morley
456.6500	462.1500	Leek
456.5250	462.0250	Leigh
456.9875	462.4875	Letchworth
453.0625	459.5625	Lincoln
456.5250	462.0250	Litchfield
453.8000	460.3000	Littlehampton
456.3500	461.8500	Liverpool
456.8250	462.3250	Liverpool
456.9500	462.4500	Liverpool
453.1000	459.6000	Llanelli
453.7750	460.2750	London
453.8500	460.3500	London
456.8750	462.3750	London
443.4750	428.9750	London N1
456.7500	462.25000	London Surrey Quays
166.7625	171.5625	London Oxford Street John Lewis
166.7625	171.5625	London Brent Cross John Lewis
453.7750	460.2750	Londonderry
453.8000	460.3000	Loughborough
456.8750	462.3750	Luton
453.4750	459.9750	Maidenhead
456.0500	461.5500	Maidstone
456.5500	462.0500	Malvern
453.9250	460.4250	Mansfield
453.9750	460.4750	March
453.1000	459.6000	Maryport
453.6500	460.1500	Matlock
456.3875	456.3875	Meadow Bank, Lothian
453.0625	459.5625	Middlesbrough
453.7750	460.2750	Milton Keynes
453.7750	460.2750	Monmouth
453.6000	460.1000	Morecambe
167.0375	171.8375	Musselborough Edinburgh
456.7000	462.2000	Neath
453.0875	459.5875	Newark
456.9500	462.4500	Newbury
456.5250	462.0250	Newcastle under Lyme

137

Base/Simplex	Portable	Location		Base/Simplex	Portable	Location
453.0625	459.5625	Newport (IoW)		456.0250	461.5250	Stalybridge
165.7875	170.5875	Newquay		453.9250	460.4250	Stamford
453.7750	460.2750	Northampton		453.7500	460.2500	Stevenage
453.8000	460.3000	Northwich		456.3875	461.8875	Stevenage
453.7500	460.2500	Ormskirk		167.0375	171.8375	Stockport
453.8000	460.3000	Oxford		440.9250	426.4250	Stockport
456.5250	462.0250	Paignton		453.1500	459.6500	Stockport
456.6000	462.1000	Paisley		453.7500	460.2500	Stoke On Trent
453.1125	459.6125	Penrith				Festival Park
166.7625	171.5625	Plymouth		440.9125	426.4125	Stourbridge
453.1750	459.6750	Plymouth		456.9250	462.4250	Stowmarket
453.7500	460.2500	Plympton		456.9500	462.4500	Stratford upon Avon
167.0375	171.8375	Pontefract		456.9250	462.4250	Sudbury
453.1000	459.6000	Poole		456.5250	462.0250	Sunderland
453.0250	459.5250	Port Talbot		456.6750	462.1750	Sutton
456.8750	462.2750	Portsmouth		456.3500	461.8500	Swanley
456.9500	462.4500	Preston		169.3000	169.3000	Swansea Morriston
453.0625	459.5625	Ramsgate		453.1500	459.6500	Swansea
453.6000	460.1000	Reading		453.7000	460.2000	Tamworth
440.9125	426.4125	Redditch		453.0625	459.5625	Teesside
453.7750	460.2750	Retford		456.7500	462.2500	Tewkesbury
456.8750	462.3750	Rochdale		453.9250	460.4250	Thame
456.8250	462.3250	Rotherham		453.8250	460.3250	Thetford
453.7500	460.2500	Royston		456.9500	462.4500	Torquay
166.7625	171.5625	Rugby		453.8250	460.3250	Troon
456.8750	462.3750	Rustington		453.7000	460.2000	Wakefield
456.0250	461.5250	Ryde (low)		453.9250	460.4250	Walsall
453.8000	460.3000	S London		453.0625	459.5625	Walthamstow
453.5000	460.0000	Salisbury		453.6000	460.1000	Warrington
453.1125	459.6125	Sandown (low)		456.8250	462.3250	Warwick
453.0875	459.5875	Scunthorpe		453.7000	460.2000	Waterlooville
456.3500	461.8500	Selsey		453.1125	459.6125	Waterlooville
453.7000	460.2000	Sevenoaks		453.0875	459.5875	Watford
453.4750	459.9750	Shanklin (low)		453.6000	460.1000	Watford
453.9250	460.4250	Sheerness		456.8250	462.3250	Wellingborough
453.0500	459.5500	Sheffield		453.1125	459.6125	Welwyn Garden City
453.5000	460.0000	Sherborne		453.0250	459.5250	Wembley
453.7500	460.2500	Shipley		453.0875	459.5875	Weston Super Mare
453.0500	459.5500	Shrewsbury		453.0625	459.5625	Whitby
453.7500	460.2500	Sittingbourne		453.7750	460.2750	Whitehaven
453.1125	459.6125	Skelmersdale		453.2000	459.7000	Wigston
453.1125	459.6125	Skipton		443.0000	428.5000	Wimbledon
456.5250	462.0250	Slough		453.8000	460.3000	Winchester
456.7000	462.2000	Solihull		456.0250	461.5250	Windsor
453.1750	459.6750	Southampton		453.2000	459.7000	Wisbech
453.9875	460.4875	Southampton		456.7500	462.2500	Witney
453.1125	459.6125	Southend On Sea		456.7000	462.2000	Woking
456.7000	462.2000	Southport		453.4750	459.9750	Wolverhampton
453.0875	459.5875	Southshields		453.7000	460.2000	Wolverhampton
456.6750	462.1750	St Albans		453.9250	460.4250	Worcester
166.7625	171.5625	St Andrews		453.7500	460.2500	Worksop
456.4000	461.9000	St Helens		456.6750	462.1750	Wrexham
456.7500	462.2500	Staines		453.8500	460.3500	York

Supermarket Frequencies

ASDA Wal-Mart
449.3125
449.4000
449.4750
461.4500

Marks and Spencer
449.3125
449.3125

449.4000
449.4750
453.6375
453.6375
460.1375
461.3375
461.3625
461.3875

461.4625	459.475/161.1125
462.2250	**Tesco**
462.2500	446.0375
Morrison	449.3125
449.3125	449.4000
449.4000	449.4750
449.4750	453.1000
Sainsbury	453.5875
449.3125	459.5750
449.4000	460.0750
449.4750	460.0875
462.0000	460.9875
Sainsbury Low Power Intercom Radio;	461.3125
453.0250/459.5250	461.3375
453.0625/459.5625	461.3500
453.0875/459.5875	461.3625
453.1125/459.6125	461.4625
453.1750/459.6750	462.0500
453.5875/460.0875	**Tesco Low Power Intercom Radio;**
453.6000/460.1000	453.0625/459.5625
456.8250/462.3250	**Tesco Voice Paging;**
Sainsbury Voice Paging;	459.2750/161.0000
459.475/161.1375	459.3750/161.0375
459.475/161.1125	459.4000/161.0625
459.475/161.0000	

Pubwatch Channels

(Note; many Shopwatch channels, given above, also 'double up' in the evening and weekends as Pubwatch channels).

Base/Simplex	Portable	Location
	158.7875	Southsea
165.1375	169.9375	Great Yarmouth
165.1625	169.9625	Bracknell
165.1625	169.9625	Carlisle
165.1625	169.9625	Kettering
165.1875	169.9875	Newquay
165.5250	170.3250	Kettering
165.5625	170.3625	Bridgwater
165.7875	170.5875	Gloucester
166.7625	171.5625	Swindon
440.3500	425.8500	Bolton
440.3750	425.8750	Wigan
440.9000	426.4000	Bolton
440.9500	426.4500	Bolton
441.0125	426.5125	Bolton
441.1750	426.6750	Tamworth
441.3750	426.8750	Wigan
441.7750	427.2750	Greater Manchester
441.9125	427.4125	Birkenhead, New Brighton, Wallasey, Wirral
442.8750	428.3750	Derby
445.8000	425.3000	Newcastle
446.1250		St Andrews
449.4000		Cheltenham
449.4000		Dundee
449.4750		Cheltenham
453.0500	459.5500	North Shields
453.0625	459.5625	Clacton on Sea
453.0625 Birmingham	459.5625	Harborne,
453.0625	459.5625	Westhoughton

Base/Simplex	Portable	Location
453.0875	459.5875	Bedford
453.1000	459.6000	Eyemouth
453.1125	459.6125	Liverpool
453.1750	459.6750	Aylesbury
453.1750	459.6750	Bristol
453.2500	459.750	Worksop
453.2500	459.7500	Swansea
453.6000	460.1000	Bexhill
453.7500	460.2500	Worcester
453.7750	460.2750	Aberdeen
453.7750	460.2750	Cardiff
453.7750	460.2750	Chelmsford
453.7750	460.2750	West London
453.7750	460.2750	Middlesborough
453.7750	460.2750	Southend On Sea
453.8000	460.3000	Atherton
453.8000	460.3000	Stone
453.8250	460.3250	Greater Manchester
453.8375	460.3375	Glossop (Glos)
453.8375	460.3375	Saltcoats
453.9250	460.4250	Blackpool
453.9250	460.4250	Cambridge
456.0250	461.5250	Ryde
456.0500	461.5500	Eastbourne
456.3250	461.7250	Colchester
456.3500	461.8500	Chesterfield
456.3500	461.8500	Stafford
456.3500	461.8500	Stowmarket
456.3875	462.4875	Weston Super Mare
456.4750	461.9750	Brighton
456.5250	462.0250	Lewes
456.6000	462.1000	Barnstable

139

SCANNERS

Base/Simplex	Portable	Location
456.6000	462.1000	Reading
456.6500	462.1500	Barrow in Furness
456.6750	462.1750	Salisbury
456.6750	462.1750	Worthing
456.7000	462.2000	Ashton under Lyne
456.7000	462.2000	Cheltenham
456.7500	462.2500	Hastings
456.7500	462.2500	Heywood
456.9250	462.4250	Littlehampton

Base/Simplex	Portable	Location
456.9500	462.4500	Guildford
456.9500	462.4500	Sunderland
456.9500	462.4500	Wigan
456.9750	462.4750	Maidstone
456.9875		Banbury
456.9875	462.4875	Borehamwood
456.9875	462.4875	Nottingham
456.9875	462.4875	Sheffield
460.4500		Swansea
461.3875		Cheltenham
462.2250		Cheltenham

Chapter 7
RT (Radio Telephony)
Procedure

English is the most internationally accepted language in radio communications, yet many communicators appear to have a language all of their own. There's a good reason for some of the codes, abbreviations and expressions that are used on the air: so that misunderstandings can be avoided. The use of a set of common expressions means that even people who speak different languages can send and receive basic messages correctly. In some cases, however (early CB being a good example), the expressions used are just part of the tradition which goes with the medium. We shall first consider some things that are common to most operators.

Phonetic alphabet

Sometimes, under difficult conditions, it may be impossible to tell what the user transmitting from another station is saying. Under such circumstances it is usual to spell out the message, coding the letters as words, using the 'phonetic alphabet':

These phonetics are widely used in callsigns. For instance, amateur station G4XYZ would be Golf Four X-Ray Yankee Zulu. Similar use of phonetics

A Alpha	H Hotel	O Oscar	W Whisky
B Bravo	I India	P Papa	X X-ray
C Charlie	J Juliet	Q Quebec	Y Yankee
D Delta	K Kilo	R Roger	Z Zulu
E Echo	L Lima	S Sierra	
F Foxtrot	M Mike	T Tango	
G Golf	N November	U Uniform	
		V Victor	

will be heard in aircraft callsigns, which are usually made up of a string of letters with the first or first two letters, denoting the country of registration. Some expressions are common to most radio users:

Roger; an almost universal expression meaning 'I understand or acknowledge receipt of your message'.

Wilco; not very common nowadays and it's mainly heard in early cinema films rather than on the air currently. It means I will comply with your instructions.

Copy; a message or part of it. For instance, the expression 'I copy you' means I am able to understand you, I hear you.

Mayday; the international call of distress where grave and imminent danger is present. The word is repeated three times and means that an emergency situation has occurred. All stations on the frequency, except that calling Mayday and that providing assistance, should observe strict radio silence.

Pan-Pan; a call indicating that assistance is required urgently but no one is in immediate danger. This may be followed some time later by a Mayday if the situation deteriorates.

Securite; a call giving safety and navigational information, typically of potential hazards.

Affirmative; yes.

Negative; no.

Time

Even within the relatively close confines of Europe many countries may be in different time zones and so a standard time system has been adopted so that complex calculations can be avoided during radio communications. Co-ordinated Universal Time (UTC) is the same time as Greenwich Mean Time (GMT). Occasionally, some radio operators will refer to UTC as 'Zulu', e.g., '1500 hours Zulu' is 3 o'clock in the afternoon. British summer time (BST) is known as 'Alpha', that is, UTC + 1 hour. Virtually all radio traffic references to time are made using the 24-hour clock system.

Amateurs

Amateurs form one of only two groups of radio users (the other is CB) who usually 'transmit blind': that is they put out calls for contact with anyone who happens to be on the same frequency or channel. Professional users, on the other hand, except in emergencies, only put out calls for specific stations. But amateurs, too, might well call up particular stations. And, even when transmitting blind, they may well specify that they only want contacts into a certain area. For instance, it is not unusual under 'lift' conditions to hear UK amateurs calling for contacts on the continent or even from a specific country. While most amateurs will happily chat to anyone who happens to be on the air many will, at times, only want to work long distances. One of the attractions of the hobby is being able, on occasions, to work not only far-flung places but also small countries where there may only be a few amateurs. Such 'catches' are a little bit like a stamp collector finding a rare stamp. Amateurs use expressions known as 'Q-codes' to abbreviate messages. Some typical Q-codes follow. Note that most can be either a statement or a question, e.g., QRP can mean 'shall I reduce my power?' or 'reduce your power', depending on the context of use.

International Q-codes

QRM; interference. This is 'man-made', such as noise from electrical equipment or other radio signals.

QRN; interference. Natural interference, such as static.

QRP; reducing transmitter power. The expression 'QRP station' means a transmitter that is always operated at very low power. Some amateurs specialise in this kind of operation.

QRT; stop sending/transmitting. A station saying "I am going QRT" usually means he is closing down.

QRZ; who is calling?

QSB; signals fading.

QSK; can I break in on your contact. Often a query from a station wanting to join in a 'net', that is, a group of amateurs passing conversation back and forth.

QSL; acknowledge receipt. A 'QSL card' is an acknowledgment card sent to conform a contact and are often collected in order to claim awards.

QSO; communicate or communication. For example 'I had a QSO with a French station'.

QSY; change frequency or channel. For example 'Let's QSY to 144.310MHz'.

QTH; location of station. Sometimes you may hear the expression 'QTH locator'. This is a grid system used by amateurs to work out the distances between each other.

There are many other Q-codes but they are rarely used by amateurs using speech for communications.

Reporting codes

Amateurs, like other radio users, have a system of reporting on the signal that they receive. The other station will usually find this information useful as it can tell him what propagation conditions are like and if his equipment is performing correctly. Like some scanners, amateur radio transceivers usually have a signal strength meter to indicate received signal strength. The lower part of the scale is usually marked from 0 to 9, above this the scale is marked in decibels (dB). The internationally recognised method of reporting on signals is known as the 'RST code': R is readability, S is signal strength and T is tone. For speech communication the T is not used as it applies only to Morse code. Amateurs will usually be heard to say something like 'you are five and nine'. That means readability 5, signal strength 9. The code follows:

Readability

R1; Unreadable
R2; Barely readable
R3; Readable with considerable difficulty
R4; Readable with practically no difficulty
R5; Perfectly readable

Signal strength

S1; Faint, barely perceptible
S2; Very weak
S3; Weak
S4; Fair
S5; Fairly good
S6; Good
S7; Moderately strong
S8; Strong
S9; Extremely strong
S9+; Meter needle exceeding S9 level of the scale
The last one is an unofficial code but often used and a corresponding measurement may be given in decibels, e.g., 'You are 20dB over 9'.

Call sign prefixes

It's possible to identify the country from which a station is transmitting by the first few letters and/or numbers of the callsign. The accompanying table gives a prefix list, so that from the callsign prefix you can identify where the station is located

Contest stations

Occasionally you may hear contests in operation. Participating stations, operated by an individual or a group, are required to make as many contacts as possible within a given space of time. A major UK-based VHF/UHF contest which you may hear on your scanner is probably the VHF National Field Day (VHF NFD) which is organised by the Radio Society of Great

Britain. This takes place every year on the first weekend in July between 3.00pm on the Saturday and 3.00pm on the Sunday. During the event the whole spectrum around plus and minus a few hundred kilohertz of 144.300MHz (2m) and 432.200MHz (70cm) comes alive with thousands of transmitting stations. Most of the transmissions are SSB utilising Upper Sideband, so will only be of interest to owners of suitably equipped scanners. However, for such owners this event usually provides an occasion to hear a lot of long distance stations. Lift conditions are normally good at this time of year and many continental stations beam their transmission towards the UK in order to take part in the contest. On HF you'll often hear a contest of some form nearly every weekend in the summer months, and often rare countries, i.e. those with very few amateurs on air, will appear during these times. Again the accompanying prefix list will help you identify these.

Amateur callsign prefixes, with associated countries

Prefix	Country	Prefix	Country
2E	England	3C0	Annobon
2D	Isle of Man	3D6	Swaziland
2I	Northern Ireland	3DA0	Swaziland
2J	Jersey	3D2	Conway Reef
2M	Scotland	3D2	Fiji Is
2U	Guernsey & Dependencies	3D2	Rotuma
2W	Wales	3D2/C	Conway Reef
3A	Monaco	3D2/F	Fiji Islands
3B6	Agalega	3D2/R	Rotuma
3B7	Agalega	3E, 3F	Panama
3B8	Mauritius	3G	Chile
3B9	Rodriguez Is	3H, 3I, 3J,	China
3C	Equatorial Guinea		

Prefix	Country
3K, 3L, 3M, 3N,	
3O, 3P, 3Q, 3R,	
3S, 3T, 3U	
3V	Tunisia
3W	Vietnam
3X	Guinea
3Y	Bouvet Is
3Y	Peter Is
3Y/B	Bouvet
3Y/P	Peter Is
3Y0, 3Y1,	
3Y2, 3Y9	Antarctic
3Z	Poland
4A, 4B, 4C	Mexico.
4D, 4E, 4F,	
4G, 4H, 4I	Philippines
4J , 4K	Azerbaijan
4J1, 4K1, 4J0,	
4K0	special licenses, foreigners
4J2, 4K2	Nakhichevan
4J4-4J9,	
4K4-4K9	Baku
4J3, 4K3	Azerbaijan territories excluding
	Baku and Nakhichevan
4L	Georgia
4M	Venezuela
4O	Republic of Montenegro
4P, 4Q,	
4R, 4S	Sri Lanka
4T	Peru
4U1ITU	Geneva
4U1UN	New York NY USA
4U1VIC	Vienna Intl.
4U1WB	The World Bank,
	Washington DC, USA
4V	Haiti
4W	East Timor.
4U1ET	United Nations club
	station within East Timor
4X	Israel
4Y	International Civil
	Aviation Organization
4Z	Israel
5A	Libya
5B	Cyprus
5C, 5D,	
5E, 5F, 5G	Morocco
5H, 5I	Tanzania
5H0	Dodoma, Singida
5J, 5K	Columbia
5L, 5M	Liberia
5N, 5O	Nigeria
5P, 5Q	Denmark
5R, 5S	Malagasy Rep
5T	Mauritania
5U	Niger
5V	Togo
5W	Western Samoa
5X	Uganda
5Y, 5Z	Kenya
6A, 6B	Egypt
6C	Syria
6D, 6E, 6F,	
6G, 6H,	

Prefix	Country
6I, 6J	Mexico
6K, 6L, 6M, 6N	South Korea
6O	Somalia
6P, 6Q,	
6R, 6S	Pakistan
6T, 6U	Sudan
6V, 6W	Senegal
6W1	Dakar
6W2	Ziguinchor
6W3	Diourbei
6W4	St. Louis
6W5	Tambacounda
6W6	Kaolack
6W7	Thies
6W8	Louga
6W9	Fatick
6X	Malagasy Rep
6Y	Jamaica
6Z	Liberia
7A, 7B, 7C,	
7D, 7E, 7F,	
7G, 7H, 7I	Indonesia
7J, 7K, 7L,	
7M, 7N	`Japan
7O	Yemen Republic
7P	Lesotho
7Q	Malawi
7R	Algeria
7S	Sweden
7T, 7U, 7V,	
7W, 7X, 7Y	Algeria
7Z	Saudi Arabia
8A, 8B, 8C,	
8D, 8E, 8F,	
8G, 8H, 8I	Indonesia
8J, 8K, 8L,	
8M, 8N Japan	(Used for Special Events)
8O	Botswana
8P	Barbados
8Q	Maldive Is
8R1	East Demerara, West Demerara
8R2	East Berbice, West Berbice
8R3	Essequibo, Essequibo Island,
	Mazaruni, Northwest, Rupununi
8S	Sweden
8T, 8U, 8V,	
8W, 8X, 8Y	India
8Z Saudi	Arabia
9A	Croatia
9B, 9C, 9D	Iran
9E, 9F	Ethiopia
9G	Ghana
9H	Malta
9I, 9J	Zambia
9K	Kuwait
9L	Sierra Leone
9M0	Spratly Is
9M2	West Malaysia
9M4	West Malaysia
9M6	East Malaysia
9M8	East Malaysia
9N	Nepal
9O, 9P, 9Q,	

Prefix	Country
9R, 9S, 9T	Zaire
9U	Burundi
9V	Singapore
9W	West/East Malaysia
9X	Rwanda
9Y, 9Z	Trinidad
A2	Botswana
A3	Tonga
A4	Oman
A5	Bhutan
A6	United Arab Emirates
A7	Qatar
A8	Liberia
A9	Bahrain
AC6	West Carolines
AH0	Mariana Is
AH1	Baker Howland
AH2	Guam
AH3	Johnston Is
AH4	Midway
AH5	Palmyra Is
AH5K	Kingman Reef
AH6	Hawaii
AH7	Hawaii
AH7K	Kure Is
AH8	American Samoa
AH9	Wake Is
AL	Alaska
AM, AN, AO	Spain
AM6	Balearic Is
AM8	Canary Is
AM9	Ceuta-Melilla
AN	Spain
AN6	Balearic Is
AN8	Canary Is
AN9	Ceuta-Melilla
AO	Spain
AO6	Balearic Is
AO8	Canary Is
AO9	Ceuta-Melilla
AP, AQ, AR, AS	Pakistan
AT, AU, AV, AW	India
AT4	Andaman Is
AT7	Laccadive Is
AU I	ndia
AU4	Andaman Is
AU7	Laccadive Is
AV	India
AV4	Andaman Is
AV7	Laccadive Is
AW	India
AW4	Andaman Is
AW7	Laccadive Is
AX	Australia
AY, AZ	Argentina
BA, BB, BC, BD, BE, BF, BG, BH, BI, BJ BK, BL, BP, BR, BS, BT, BU, BW, BY, BZ	China

Prefix	Country
BS7	Scarborough Reef
BM, BO, BQ, BV	Taiwan
BO2	Kinmem Island
BQ9	Pratus Island
BV	Taiwan
BV9A	Penhu Island
BV9O	Orchid Island
BV9G	Green Island
BV9P	Pratas Island
C2	Nauru
C3	Andorra
C4	Cyprus
C5	Gambia
C6	Bahamas
C8, C9	Mozambique
CA, CB, CC, CD, CE	Chile
CE9	Antarctica
CE0A	Easter Island
CE0E	Easter Island
CE0F	Easter Island
CE0I	Juan Fernandez
CE0X	San Felix/San Ambrosio
CE0Y	Easter Island
CE0Z	Juan Fernandez (
CF, CG, CH, CI, CJ, CK	Canada
CL, CM	Cuba
CN	Morocco
CO	Cuba
CP	Bolivia
CQ	Special Prefixes (CQ3 and CQ9 reserved for Madeira)
CR	Portugal Special Prefixes
CR3 and CR9	reserved for Madeira
CS	Portugal Special Prefixes
CS3 and CS9	reserved for Madeira
CT	Portugal
CT9	Madeira Isl.
CU1	Santa Maria Island
CU2	Sao Miguel Island
CU3	Terceira Island
CU4	Graciosa Island
CU5	Sao Jorge Island
CU6	Pico Island
CU7	Faial Island
CU8	Flores Island
CU9	Corvo Island
CV, CW, CX	Uruguay
CY, CZ	Canada
CY0	Sable Is
CY9	St Paul Is
D2, D3	Angola
D4	Cape Verde
D5	Liberia
D6	Comoros
D7, D8, D9	South Korea

Prefix	Country
DA, DJ, DK,	
DL, DM	Germany
DS, DT, DU ,	
DV, DW, DX,	
DY, DZ	Philippines
E2	Thailand
E3	Eritrea
E4	Palestine
E7	Bosnia Herzegovina
EB	Spain
EB6	Balearic Is
EB8	Canary Is
EB9	Ceuta
EC	Spain
EC6	Balearic Is
EC8	Canary Is
EC9	Ceuta
EC9	Melilla
ED	Spain
ED6	Balearic Is
ED8	Canary Is
ED9	Ceuta
ED9	Melilla
EE	Spain
EE6	Balearic Is
EE8	Canary Is
EE9	Ceuta
EF	Spain
EF6	Balearic Is
EF8	Canary Is
EF9	Ceuta
EG	Spain
EG6	Balearic Is
EG8	Canary Is
EG9	Ceuta
EH	Spain
EH6	Balearic Is
EH8	Canary Is
EH9	Ceuta
EI, EJ	Ireland
EK	Armenia
EL	Liberia
EM, EN, EO	Ukraine
EP, EQ	Iran
ER	Moldova
ES	Estonia
ES0	West coast islands
ET	Ethiopia
EU, EV, EW	Belarus
EX	Kyrgyzstan
EY	Tadjikistan
EZ	Turkmenistan
F	France
FG	Guadeloupe
FH	Mayotte
FJ St.	Barthelemy (St. Barts),
FK	Chesterfield Isle
FK	New Caledonia
FM	Martinique
FO	Australs
FO	Clipperton
FO	Tahiti
FO	Marquesas Isle
FO/C	Clipperton

Prefix	Country
FP	St Pierre Miquelon
FR	Glorioso
FR	Juan De Nova
FR	Reunion
FR	Tromelin
FR/G	Glorioso
FR/J	Juan De Nova
FR/T	Tromelin
FS	St Martin
FT0W	Crozet
FT0X	Kerguelen Is
FT0Z	Amsterdam Paul
FT2W	Crozet
FT2X	Kerguelen Is
FT2Z	Amsterdam Paul
FT5W	Crozet
FT5X	Kerguelen Is
FT5Z	Amsterdam Paul
FT8W	Crozet
FT8X	Kerguelen Is
FT8Y	Antarctica
FT8Z	Amsterdam Paul
FU	France
FV	France
FW	Wallis Is
FY	French Guiana
FX	French amateur satellites
G	England
GB	Special Event Station in the UK,
	Channel Is or Isle of Man
GC	Wales
GD	Isle of Man
GH	Jersey (Clubs)
GI	Northern Ireland
GJ	Jersey
GM	Scotland
GN	Northern Ireland
GP	Guernsey & Dependencies (Clubs)
GS	Scotland (Clubs)
GT	Isle Of Man (Clubs)
GU	Guernsey & Dependencies
GW	Wales
GX	England (Clubs
H2	Cyprus
H3	Panama
H4	Solomon Is
H40	Temotu
H6, H7	Nicaragua
H8, H9	Panama
HA	Hungary
HB	Switzerland
HB0	Liechtenstein
HC, HD	Ecuador
HC8	Galapagos
HD8	Galapagos
HE	Switzerland
HF	Poland
HF0 So	Shetland
HF0	Antarctica
HG	Hungary
HH	Haiti
HI	Dominican Rep
HJ, HK	Columbia
HK0/A	San Andres Is

149

Prefix	Country	Prefix	Country
HK0/M	Malpelo Is	KP3	Puerto Rico
HL	South Korea	KP4	Puerto Rico
HM	North Korea	KP5	Desecheo Is
HN	Iraq	LA, LB, LC,	
HO, HP,		LD, LE, LF,	
HQ, HR	Honduras	LG, LH, LI,	
HS	Thailand	LJ, LK, LL,	
HT	Nicaragua	LM, LN	Norway
HU El	Salvador	L2, L3, L4,	
HV	Vatican City	L5, L6, L7,	
HW, HX, HY	France	L8, L9	Argentina
HZ	Saudi Arabia	LX	Luxembourg
I	Italy	LY	Lithuania
IT9, IW9	Sicily	LZ	Bulgaria
IS0, IW0		LZ0	Antarctica
UAA-ZZZ	Sardinia	M	England
IA1, IP1	Liguria	MC	Wales (Clubs)
IA5	Toscana	MD	Isle of Man
IB0	Lazio	MH	Jersey (Clubs)
IC8	Campania	MI	Northern Ireland
ID8	Calabria Islands Dino and Cirella	MJ	Jersey
ID9,		MM	Scotland
IE9, IF9	Sicily	MN	Northern Ireland
IG9, IH9	Sicily	MP	Guernsey & Dependencies
IG9	Lampedusa,	MS	Scotland (Clubs)
IH9	Pantelleria	MT	Isle Of Man (Clubs)
IJ7, IL7	Puglia	MU	Guernsey & Dependencies
IM0	Sardinia	MW	Wales
IL3	Veneto	MX	England
IA, IR, IY	Antarctica	N	United States
J2	Djibouti	NC6	East Carolines
J3	Grenada	NC6	West Carolines
J4	Greece	NC6/E	East Caroline
J45	Dodecanese	NC6/W	West Carolines
J49	Crete	NH0	Mariana Is
J5	Guinea Bissau	NH1	Baker Howland
J6	St Lucia	NH2	Guam
J7	Dominica	NH3	Johnston Is
J8	St Vincent	NH4	Midway Is
JT, JU, JV	Mongolia	NH5	Palmyra Is
JW	Svalbard Is	NH5K	Kingman Reef
JX	Jan Mayen	NH6	Hawaii
JY	Jordan	NH7	Hawaii
JZ	Indonesia	NH7K	Kure Is
K	United States	NH8	American Samoa
KC4	Antarctica Bryd	NH9	Wake Is
KC4	Antarctica McMurdo	NL	Alaska
KC4	Antarctica	NP1	Navassa Is
KG4	USA or Guantanamo Bay	NP2	Virgin Is
KH0	Mariana Is KH1 Baker Howland	NP3	Puerto Rico
KH2	Guam	NP4	Puerto Rico
KH3	Johnston Is	NP5	Desecheo Is
KH4	Midway Is	OA, OB, OC	Peru
KH5	Palmyra Is	OA0	Antarctica
KH5K	Kingman Reef	OD	Lebanon
KH6	Hawaii	OE	Austria
KH7	Hawaii	OF, OG,	
KH7K	Kure Is	OH, OI, OJ	Finland
KH8	American Samoa	OH0	Aland Is
KH8	Swains Island	OH0M	Market Reef
KH9	Wake Is	OJ0	Market Reef
KL	Alaska	OK, OL	Czech Republic
KP1	Navassa Is	OM	Slovakia
KP2	Virgin Is		

Prefix	Country
ON, OO, OP,	
OQ, OR,	
OS, OT	Belgium
OU, OV,	
OW, OX, OY,	
OZ	Denmark
OX	Greenland
OY	Faroe Is
P2	Papua
P3	Cyprus
P30	Cyprus
P36	Cyprus
P4	Aruba
P5, P6, P7,	
P8, P9	North Korea
PA, PB, PC,	
PD, PE, PF,	
PG, PH, PI	Netherlands
PJ	Netherlands Antilles
PJ	St Maarten, France
PJ2	Curacao, Netherlands Antilles
PJ4	Bonaire, Netherlands Antilles
PJ5	St. Eustatius
PJ6	Saba
PJ7	Sint Maarten
PJ8	Sint Maarten
PJ9	Netherlands Antilles
PK, PL,	
PM, PN, PO	Indonesia
PP, PQ, PR,	
PS, PT, PU,	
PV, PW,	
PX, PY	Brazil
PX-	Brazil Citizen Band Assignments
PY0A-	Abrolhos Is
PY0F-	Fernando de Noronha Is.
PY0M-	Martin Vaz Is.
PY0R-	Atol das Rocas Is.
PY0S-	Sao Pedro & Sao Paulo
PY0T-	Trindade Is.
PZ	Suriname
R	Russian Federation
S0	Western Sahara
S1A	Principality of Sealand (Unofficial prefix) 6 miles off the eastern shores of Britain
S2, S3	Bangladesh
S5	Slovenia
S6	Singapore
S7	Seychelles
S8	South Africa
S9	Sao Tome
SA, SB, SC,	
SD, SE, SF,	
SG, SH, SI,	
SJ, SK,	
SL, SM	Sweden
SN, SO, SP,	
SQ, SR	Poland
SS	Egypt
ST	Sudan
SU	Egypt
SV, SW, SX, SY, SZ	Greece
SV/A	Mount Athos
SV5	Dodecanese
SV9	Crete

Prefix	Country
T2	Tuvalu
T3	Kiribati
T4	Cuba
T5	Somalia
T6	Afghanistan
T7	San Marino
T88	Belau
T9	Bosnia Herzegovina
TA	Turkey
TD	Guatemala
TE	Costa Rica
TF	Iceland
TG	Guatemala
TH	France
TI	Costa Rica
TI9	Cocos Is
TJ	Cameroon
TK	Corsica
TL	Central Africa Rep
TM	France/Europe (Outside France) mostly used during contests.
TN	Congo
TO, TO,	
TP, TQ	France
TR	Gabon
TS	Tunisia
TT	Chad
TU	Ivory Coast
TV, TW, TX	France
TX	France, TX is for TOM (Territoires d'outre-mer)
TX	Chesterfield Isle
TY	Benin
TZ	Mali
UA-UI	Russian Federation
UJ, UK,	
UL, UM	Uzbekistan
UN, UO, UP,	
UQ	Kazakhstan
UR, US, UT,	
UU, UV,	
UW, UX, UY,	
UZ	Ukraine
V2	Antigua
V3	Belize
V4	St Kitts
V5	Namibia
V6	Fed Micronesia
V7	Marshall Is
V85	Brunei
VA VB VC	
VD VE	
VF VG	Canada
VY0	Nunavut
VH VI VJ	
VK	Australia
VK9/C	Cocos Keeling
VK9/H	Lord Howe
VK9/M	Mellish Reef
VK9/N	Norfolk Is
VK9R	Rowley Shoals Atolls
VK9/W	Willis Is
VK9X	Christmas Is
VK0	Heard Is
VK0	Macquarie Is
VK0	Antarctica
VP2E	Anguilla

151

VP2M	Montserrat
VP2V	Brit Virgin Isles
VP5	Turks Caicos
VP6	Pitcairn
VP6D	Ducie Island
VP8/G	So Georgia
VP8/O	So Orkney
VP8/SA	So Sandwich
VP8/SH	So Shetland
VP8	Antarctica
VP9	Bermuda
VQ9	Chagos
VR2	Hong Kong
VT, VU, VV,	
VW	India
VU7	Laccadive Is
VU7	Andaman & Nicobar Isles
VU	Antarctica
VX, VY	Canada
VY1	Yukon
VY2	Prince Edward Is
VY9	Prince Edward Is
VY9	Government of Canada
VZ	Australia
W	United States

Special event stations

Occasionally you may hear 'special event stations' which are usually operated by a group of amateurs such as an amateur radio club. They are granted a special one-off callsign to celebrate special events such as a country fair or a scout Jamboree-on-the-air. Callsigns are often granted to have some significance to the event. For instance, amateurs operating from the Totnes Agricultural Fair might use the callsign GB2TAF. The GB prefix is the normal one used for special event stations although many stations use the GX prefix to allow transmissions of a short greetings message by visitors to the station.

Repeaters

Details of how repeaters work were outlined in Chapter 2. Repeaters in the 6 metre, 2 metre, 70 cm and 23cm bands operate in FM mode and so can be received on any scanner that covers the bands. They are often recognised by a periodic transmission of Morse code containing the callsign of the repeater. Amateurs often call blind on repeaters and you may well hear the expression 'This is G1ABC listening through GB3SH'. That means that amateur station G1ABC has accessed the repeater with callsign GB3SH and is awaiting any replies.

Marine

The international VHF marine band as we saw in Chapter 5 is channelised. The usual procedure at commencement of a normal speech transmission on the band is to first put out a call on channel 16: the calling channel, requesting contact with a particular station. When that station replies, both then move to a 'working channel'. This method of operation means that at any time there might be hundreds of stations listening to channel 16, and so if any boat or ship needs help someone is usually bound to hear the call. A further advantage is that it enables shore stations to make general broadcasts to ships informing them that weather and safety warnings are about to be transmitted on a given channel. This method of initial calls is however being replaced by the GMDSS (Global Maritime Distress and Safety System) where initial transmissions employ a data transmission on channel 70 which contains information as to the nature of the call, and where appropriate designates a speech 'working' channel for subsequent communications. However, the UK coastguard intend to also maintain an aural listening watch for some years on channel 16, and the owners of many pleasure craft will continue to use just speech communication. Vessels licensed for marine RT and GMDSS are given a callsign comprising letters, numbers or both, depending on where the ship is registered. Generally the official callsign is only used when establishing link calls. For other contacts the vessel will only usually give the ship or boat's name.

Securite

Pronounced 'securitay' this word, repeated three times, precedes any broadcast transmission where there is reference to safety. Again, broadcasts telling ships that there is a securite message will be initially transmitted on channel 16, and where appropriate will give the channel to move to for the details. Securite broadcasts usually concern 'navigational warnings'. Typically they might inform vessels that a certain beacon or lighthouse is out of action, or they might warn of floating obstructions such as cargo washed off a ship's deck or a capsized vessel. Some shore stations have the task of making regular broadcasts in busy shipping areas where there may be a need to pass frequent safety messages. Typical is Cherbourg Radio (channel 11) which transmits safety information every half hour for the southern part of the English Channel - possibly the busiest shipping zone in the world.

Weather

All coastal stations transmit regular weather forecasts and gale warnings. Again, forewarning of a weather forecast or gale warning will usually be made on channel 16 and the station will say which working channels will carry the forecast. More localised forecasts are made by some ports, using the same procedure as the coastal station. Normally such forecasts are broadcast on the regular port operations channel.

Port operations

So far we have looked largely at the kind of transmissions and broadcasts that are from coastal stations covering a wide area. However, the marine VHF band is also used for other kinds of contact, in particular port operations. Many ports are busy places and some have traffic handling facilities almost as sophisticated as airports. Typical radio traffic concerns departure and arrival of ships, ferries and pleasure craft. Port controllers, for example, may have to hold some ships offshore until other ships have left and made space for them. They may be contacted by yachts wanting mooring spaces in marinas. Other tasks involve liaison with bodies such as customs and immigration officers.

Marina Channels

There are channels specifically set aside for pleasure craft to communicate with marinas, for example to arrange berthing. These are channel 'M' (this usually being channel 37 on suitably-equipped marine radios), and increasingly channel 80 is also used as a marina channel to accommodate foreign craft whose radios may not have channel 'M'. A further channel specific to the UK is channel M2, which is typically used by yacht clubs, and when an organised yacht race event is taking place this is the channel where much activity takes place.

Ship-to-ship

Several channels are set aside for ship-to-ship use. These are used for a variety of purposes, such as trawlermen discussing their catches are.

Emergencies

Britain has arguably the best marine emergency services in the world; its tradition as a seafaring nation is probably responsible for this. The waters around the British Isles are covered by lifeboat and coastguard stations, and support to these services comes from the Air Force and Royal Navy. All services are on call to assist with emergencies at sea. The first warning of an emergency concerning imminent danger to life will come with a Mayday call. This is internationally recognised and will either be made with a digital call on GMDSS or a speech call made on channel 16. Normally, a coastal station or port will receive the call and put the emergency services into action. Where coastal stations are out of range, a ship may well respond to the Mayday. During this time all traffic, other than emergency traffic, is supposed to cease on channel 16. No two emergencies are the same. In some instances it may just be a 'Pan Pan' or 'Securite' message from a small vessel lost in fog and worried about running onto rocks. In such cases coastal stations might be able to offer position fixes by taking bearings on the transmissions of the vessel in danger by pinpointing the vessel's position. At the other end of the scale, the emergency might be a ship sinking in a storm.

Aviation

The world of aviation is the winner when we come to judge it in terms of the number of expressions and jargon. However, little of it is trivial. Aircraft crew cope with a variety of complex situations and may well be flying in and out of countries where air traffic controllers have little if any understanding of the English language. Like the marine band the aircraft band is channelised, but the channels are referred to by their actual frequency and not by a channel number. The procedure for contacting a station is also very different. There is no common calling channel: a pilot wanting to call a ground station simply looks up the frequency and calls on it.

Aircraft Callsigns

To the uninitiated, listening-in on the air bands can often be a frustrating experience not only because of the high level of jargon used by pilots and ground stations but also by the bewildering number of different callsigns that airlines and operators use. Callsigns will in fact fall into one of two categories. The first is the prefix that denotes the country or origin; the second is a self-assigned name registered with the authorities in the country of registration. One of these options will be used by all civil aircraft. Military aircraft using civilian airways might also use a code name. Typical examples being "Reach" used by the United States Air Force and "Ascot" used by the British Royal Air Force Transport Group. First the country prefix. The accompanying table shows the letters and numbers association with the registration system for any particular country, but do be aware that it is common practice to only give the full callsign on the first contact with a ground station. From then on the last two letters or digits of the callsign are all that are used. These registration prefixes are normally used by light aircraft, those that are privately owned and the smaller commercial operators although there are instances where major airlines will use these as well.

International Aircraft callsign and registration prefixes

Prefix	Country	Prefix	Country
A2	Botswana	OY	Denmark
A3	Tonga	P	North Korea
A5	Bhutan	P2 P	apua New Guinea
A6	United Arab Emirates	PH	Netherlands
A7	Qatar	PK	Indonesia
A9	Bahrain	PJ	Netherlands Antilles
AP	Pakistan	PP/PT	Brazil
B	China/Taiwan	PZ	Surinam
CF	Canada	RDPL	Laos

CG	Canada	RP	Phillipines
C2	Nauru	S2	Bangladesh
C3	Andorra	S7	Seychelles
C5	Gambia	S9	Sao Tome
C6	Bahamas	SE	Sweden
C9	Mozambique	SP	Poland
CC	Chile	ST	Sudan
CN	Morocco	SU	Egypt
CP	Bolivia	SX	Greece
CR	Portuguese Overseas Terr.	T2	Tuvalu
CS	Portugal	T3	Kiribati
CU	Cuba	T7	San Marino
CX	Uruguay	TC	Turkey
D	Germany	TF	Iceland
D2	Angola	TG	Guatemala
D4	Cape Verde Islands	TI	Costa Rica
D6	Comores Is.	TJ	Cameroon
DQ	Fiji	TL	Central Africa
EC	Spain	TR	Gabon
EI/EJ	Eire	TS	Tunisia
EL	Liberia	TT	Chad
EP	Iran	TU	Ivory Coast
ET	Ethiopia	TY	Benin
F	France & French Terr.	TZ	Mali
G	Great Britain	V2	Antigua
H4	Solomon Islands	V3	Belize
HA	Hungarian Peoples Rep.	V8	Brunei
HB	Switzerland & Liechtenstein	VH	Australia
HC	Ecuador	VN	Vietnam
HH	Haiti	VPF	Falkland Islands
HI	Dominican Republic	VPLKA	St Kitts Nevis
HK	Columbia	VPLLZ	St Kitts Nevis
HL	South Korea	VPLMA	Montserrat
HP	Panama	VPLUZ	Montserrat
HR	Honduras	VPLVA	Virgin Islands
HS	Thailand	VPLZZ	Virgin Islands
HZ	Saudi Arabia	VQT	Turks & Caicos Is.
I	Italy	VRB	Bermuda
J2	Djibouti	VRC	Cayman Islands
J3	Grenada	VRH	Hong Kong
J5	Guinea Bissau	VT	India
J6	St Lucia	XA/XB	Mexico
J7	Dominica	XC	Mexico
J8	St Vincent	XT	Burkina Faso
JA	Japan	XU	Kampuchea
JY	Jordan	XY/XZ	Burma
LN	Norway	YA	Afghanistan
LQ/LV	Argentina	YI	Iraq
LX	Luxembourg	YJ	Vanuatu
LZ	Bulgaria	YK	Syria
MI	Marshall Islands	YN	Nicaragua
N	USA	YR	Romania
OB	Peru	YS	El Salvador
OD	Lebanon	YV	Venezuela
OE	Austria	Z	Zimbabwe
OH	Finland	ZA	Albania
OK	Czech Republic	ZK	New Zealand
OM	Slovak Republic	ZP	Paraguay
OO	Belgium	ZS	South Africa
3A	Monaco	6Y	Jamaica
3B	Mauritius	7O	Yemen
3C	Equatorial Guinea	7P	Lesotho
3D	Swaziland	7Q	Malawi
3X	Guinea	7T	Algeria
4R	Sri Lanka	8P	Barbados
4X	Israel	8Q	Maldives

5A	Libya	8R	Guyana
5B	Cyprus	9G	Ghana
5H	Tanzania	9H	Malta
5N	Nigeria	9J	Zambia
5R	Madagascar	9K	Kuwait
5T	Mauritania	9L	Sierra Leone
5U	Niger	9M	Malaysia
5V	Togo	9N	Nepal
5W	Polynesia	9Q	Congo
5X	Uganda	9U	Burundi
5Y	Kenya	9V	Singapore
6O	Somalia	9XR	Rwanda
6V/6W	Senegal	9Y	Trinidad & Tobago

Callsigns

If the full callsign consists of letters only then you can be almost certain that the callsign being used is the standard country prefix followed by the aircraft registration. The vast majority of countries use two, three or four letter groups after the country prefix but there are a few exceptions and the notable ones are:

United States of America; N followed by numbers or a mix of numbers and letters.

Japan; JA followed by a 4 digit number.

Venezuela; YV followed by a 3 digit number then suffixed with a single letter.

China/Taiwan; B followed by a 3 or 4 digit number.

Cuba; CU-T followed by a 4 digit number.

Columbia; HK followed by a 4 digit number suffixed with X.

Korea; HL followed by a 4 digit number.

All civilian aircraft have a registration. Generally, privately owned light aircraft or those operated as air taxis will use their registration as their radio callsign. Normal procedure for making contact with the ground station will be to give the full callsign. The controller will reply, perhaps, referring to the aircraft by the full callsign, in which case the pilot will again, when transmitting, use the full callsign. At some stage though, for the sake of brevity, the approach controller will just use the last two letters and from then on the pilot will do the same. Larger aircraft, such as those used on regular passenger carrying routes, may use the same type of callsign, or a special callsign based on the airline's name and typically may also use the flight number for that service. Again the ground controller will probably, at some stage, abbreviate this and just use the number: from then on the aircrew will do the same. Numbers preceded by the word 'Ascot' denote the callsign of a British military aircraft flying on a civilian route. The USAF equivalent is the pre-fix 'Mac'.

Landing instructions

The first contact an aircraft will have with an airfield is usually on the approach frequency. After transmitting on the frequency and identifying the aircraft, the pilot usually gives aircraft position and altitude. The approach controller then transmits information relating to airfield barometric pressure (QFE), the wind direction and speed, the runway in use (runways are always identified by the compass heading needed to land on them), and details of other aircraft in the landing pattern or about to take off. Temperature and visibility in kilometres may be also given. If the weather is bad the RVR (runway visual range) may be referred to. The pilot needs to know the QFE (sometimes just called the 'fox echo') so that the aircraft's altimeter may be set so that it will read zero feet at runway level. Other information such as runway state (if affected by rain, ice or snow) may be transmitted, followed by instructions to remain at present altitude or start descending to circuit height (often about 1000 feet). Some of this information the pilot will repeat back. As the aircraft gets closer to the airport there will come a stage where the approach controller instructs the pilot to change to the tower frequency. The pilot always repeats the frequency to be changed to: this is standard procedure when changing frequency at any point in a flight. Now the pilot calls the tower and again will give his position and altitude. The aircraft may be making a straight-in approach, that is, arriving at the airfield in line with the runway or he may be 'joining the circuit'. The circuit is an imaginary path around the airfield in the form of a racetrack. It can be in a left-hand or right-hand direction and, once joined, the pilot will report at various stages such as downwind leg, base leg and finals. Finals occur at a given distance from the runway and the pilot will always tell the controller when the aircraft is one mile out. Throughout this stage of the flight the pilot will be given various instructions and updated QFE, wind speed and direction information. At any stage of the approach the pilot may be told to divert course because the controller cannot yet fit him in with other traffic. The instruction may be to briefly orbit over a given position, or to fly out further to a given point and then re-join the landing pattern. Once on the ground the pilot may be told to change frequency yet again (particularly at larger airfields), this time to speak to the ground handler. Here, instructions on which taxiways to use and where to park the aircraft will be given.

Airline call signs

Callsign	Operator	Country
Aceforce	NATO Command	Europe Mil
Ace Air	Air Cargo Express	USA
Actair	Air Charter and Travel	UK
Aeradio	International Aeradio	UK
Aero	United States Army	US Mil
Aero Lloyd	AeroLloyd	Germany
Aeroflot	Aeroflot	Russia

Callsign	Operator	Country
Aeromar Com		
Aeromaritime d'Affretement		France
Aeromaritime		
Aeromaritime		France
Aeromexico	Aeromexico	Mexico
Aeronaut	Cranfield Institute of Technology	UK Gov

Callsign	Operator	Country
Aeroperu	Aeroperu	Peru
Aeroswede	Syd Aero	Sweden
Afro	Affretair	Zimbabwe
AirafricAir	Afrique	Ivory Coast
Air America	Air America	USA
Air Atlantis	Air Atlantis	Portugal
Air Belgium	Air Belgium	Belgium
Air BVI	Air BVI	BVI
Air Canada	Air Canada	Canada
Aircal	Air Caledonie	France
Air Falcon	Europe Falcon Service	France
Air Ferry	British Air Ferries	UK
Air Force One	US President	US Mil
Air Force Two	US Vice-President	US Mil
Air France	Air France	France
Air Freighter	Aeron International	USA
Air Hong Kong	Air Hong Kong	Hong Kong
Air India	Air India	India
Air Lanka	Air Lanka	Sri Lanka
Air London	Air London	UK
Air Mauritius	Air Mauritius	Mauritius
Airmil	Spanish Air Force	Spain
Air Portugal	Air Portugal	Portugal
Air Rwanda	Air Rwanda	Rwanda
Air Services	Austrian Air Services	Austria
Air Tara	Air Tara	Eire
Airtax	Birmingham Aviation	UK
Air Zimbabwe	Air Zimbabwe	Air Zimbabwe
Airafric Coast	Air Afrique	Ivory
Airbiz	Maersk Commuter	Denmark
Airbridge	Air Bridge Carriers	UK
Aircargo	Intavia	UK
Airgo	Airgo	UK
Airmove	Skywork	UK
Airnav	Air Navigation and Trading	UK
Alisarda	Alisarda	Italy
Alitali	Alitalia	Italy
All Nippon	All Nippon	Japan
American	American Airlines	USA
Amtran	American Trans-Air	USA
Anglo	Anglo Cargo	UK
Argentine	Aerolineas Argentinas	Argentina
Armyair	Army Air Corps	UK Mil
Ascot	RAF 1 Group Air Transport	UK Mil
Aspro	Intereuropean Airways	UK
Atlantic	Air Atlantique	UK
Augusta	Augusta Airways	Australia
Austrian	Austrian Airlines	Austria
Aviaco	Aviaco	Spain
Avianca	Avianca	Colombia
Ayline	Aurigny Air Services	UK
Backer	British Charter	UK
Bafair	Belgian Air Force	Belgium
Bafjet UK	British Air Ferries Business Jets	
Bahrain One	The Amiri Flight	Bahrain
Bailair	Balair	Switzerland
Balkan	Balkan Bulgarian Airlines	Bulgaria
Batman	Ratioflug	Germany
Bangladesh	Bangladesh Biman	Bangladesh
Beaupair	Aviation Beauport	UK
Beeline	Biggin Hill Executive	UK
Birmex	Birmingham European	UK
Biztravel	Business Air Travel	UK
Beatours	British Airtours	UK
Blackbox	Bedford Royal Aircraft Estab.	UK Gov
Blackburn	British Aerospace (Scampton)	UK
Bluebird	Finnaviation	Finland
Bodensee	Delta Air	Germany
Botswana	Air Botswana	Botswana
Braethens	Braethens SAFE	Norway
Bristol	British Aerospace (Bristol)	UK
Britannia	Britannia Airways	UK
Britanny	Brit Air	France
British Island	British Island Airways	UK
Brunei	Royal Brunei Airlines	Brunei
Busy Bee	Busy Bee	Norway
Camair	Cameroon Airlines	Cameroon
Canada	Worldways Canadian	Canada
Canadian	Canadian Airlines	Canada
Canforce	Canadian Air Force	Canada
Cargo Freighters	Safair	S. Africa
Cargolux	Cargolux	Lux'bourg
Cathay	Cathay Pacific Airways	H.K.
Cayman	Cayman Airways	Cayman Islands
Cedar Jet	Middle East Airlines	Lebanon
Chad	Chad Air Services	Chad
Channex	Channel Express	UK
China	CAAC	China
City	KLM City Hopper	Holland
Clansman	Airwork Limited	UK
Conair	Conair	Denmark
Condor	Condor Flugdienst	Germany
Contactair	Contactair	Germany
Continental	Continental Airlines	USA
Corsair	Corsair	France
Crossair	Crossair	Switzerland
Cubana	Cubana	Cuba
Cyprus	Cyprus Airways	Cyprus
Dantax	Aalborg Airtaxi	Belgium
Databird	Air Nigeria	Nigeria
Dash	Air Atlantic	UK
Delta	Delta Air Lines	USA
Deltair	Delta Air Transport	Belgium
DLT	DLT	Germany
Dominair	Aerolineas Dominicanes	Dominican Rep.
Dragon	Welsh Airways	UK
Dynamite	Dynamic Air	Holland
Easyjet	Easyjet	UK
Egyptair	Egyptair	Egypt
El Al	El Al	Israel

159

Callsign	Operator	Country
Elite	Air 3000	Canada
Emery	Emery Worldwide	USA
Emirates	Emirate Airlines	UAE
Espania	CTA Espania	Spain
Ethiopian	Ethiopian Airlines	Ethiopia
Euralair	Euralair	France
Euroair	Euroair Transport	UK
Eurotrans	European Air Transport	Belgium
Evergreen	Evergreen International	USA
Excalibur	Air Exel	UK
Executive	Extra Executive	
	Transport	Germany
Express	Federal Express	USA
Falcon Jet	Falcon Jet Centre	UK
Ferranti	Ferranti Ltd	UK
Finnair	Finnair	Finland
Flamingo	Nurnberger Flugdienst	Germany
Food	Food Brokers Limited	UK
Fordair	Ford Motor Company	UK
Foyl	Air Foyle	UK
Fred Olsen	Fred Olsen	
	Air Transport	Norway
Gatwick Air	Gatwick Air Taxis	UK
Gauntlet	Boscombe Down (MOD)	UK Mil
German Cargo	German Cargo	Germany
Germania	Germania	Germany
Ghana	Ghana Airways	Ghana
Gibair	GB Airways Limited	UK
Golf November	Air Gabon	Gabon
Granite	Business Air	UK
Greenlandair	Gronlandsfly	Denmark
Gulf Air	Gulf Air	Oman
Guyair	Guyana Airways	Guyana
Hapag-Lloyd	Hapag-Lloyd	Germany
Hatair	Hatfield Executive	
	Aviation	UK
Hawker	British Aerospace	
	(Dunsfold)	UK
Hunting	Hunting Surveys	UK
Iberian	Iberia	Spain
Iceair	Icelandair	Iceland
Indonesian	Garuda Indonesian	
	Airways	Indonesia
Interflug	Interflug	Germany
Iranair	Iran Air	Iran
Iraqi	Iraqi Airways	Iraq
Janus	Janus Airways	UK
Japanair	Japan Air Lines	Japan
Jetset	Air 2000	UK
Joker	Germania	Germany
Jordanian	Royal Jordanian Airline	Jordan
KLM	KLM Royal	
	Dutch Airlines	Holland
Karair	Kar-Air	Finland
Kenya	Kenya Airlines	Kenya
Kestrel	Airtours	UK
Kilo Mike	Air Malta	Malta
Kilroe	Air Kilro	UK
Kittyhawk	Queen's Flights	UK Mil
Koreanair	Korean Air	Korea

Callsign	Operator	Country
Kuwaiti	Kuwait Airways	Kuwait
Leopard	Queen's Flights	UK Mil
Libair	Libyan Arab Airlines	Libya
Lion	British International	
	Helicopters	UK
Lovo	Lovaux Limited	UK
Lufthansa	Lufthansa	Germany
Luxair	Luxair	Lux'bourg
Macline	McAlpine Aviation Ltd	UK
Madair	Air Madagascar	Madagascar
Maerskair	Maersk Air	Denmark
Malawi	Air Malawi	Malawi
Malaysian	Malaysian Airlines	
	System	Malaya
Malev	Malev	Hungary
Mamair	Marine and Aviation	
	Management	UK
Mann	Alan Mann Helicopters	UK
Marocair	Royal Air Maroc	Morocco
Martinair	Martinair	Holland
Mediterranean	Mediterranean Express	UK
Merlin	Rolls Royce Military	UK
Metman	Meteorological Research	
	Flight	UK Gov
Metro	Bohnstedt Petersen	
	Aviation	Denmark
Midas	Milford Docks	
	Air Services	UK
Midland	British Midland Airways	UK
Midwing	Airborne of Sweden	Sweden
Mike Romeo	Air Mauritania	Mauritania
Minair	C.A.A. Flying Unit	UK Gov
Minerve	Minerve	France
Monarch	Monarch Airlines	UK
Nationair	Nation Air	Canada
National	Airmore Aviation	UK
Navy	Royal Navy	UK Mil
Neatax	Northern Executive	
	Aviation	UK
Netherlands	Royal Netherlands	
	Air Force	Holland
Netherlines	Netherlines	Holland
New Z'land	Air New Zealand	New Z'land
Newpin	British Aerospace	
	(Hawarden)	UK
Nigerian	Nigerian Airways	Nigeria
Nightflight	Night Flight	UK
Norseman	Norsk Air	Norway
Northair	Northern Air Taxis	UK
Northwest	Northwest Orient	USA
November Lima	Air Liberia	Liberia
November Papa	Heavylift Cargo Airlines	UK
Nugget	Farnborough Royal	
	Aircraft Estab.	UK Gov
Olympic	Olympic Airways	Greece
Orange	Air Holland	Holland
Orion	Orion Airways	UK
Overnight	Russow Aviation	Germany
Palmair	Palmair	UK

Callsign	Operator	Country
Pakistan	Pakistan International	Pakistan
Para	Army Parachute Centre	UK Mil
Paraguaya	Lineas Aereas Paraguayas (LAP)	Paraguay
Partnair	Partnair	Norway
Pearl	Oriental Pearl Airways	UK
Philair	Philips Aviation Services	Holland
Philippine	Philippine Airlines	Philippines
Plum	PLM Helicopters	UK
Police	Police Aviation Services	UK
Pollot	Polski Linie Lotnicze (LOT)	Poland
Port	Skyworld Airlines	USA
Puma	Phoenix Aviation	UK
Quantas	Quantas	Australia
Quebec	Tango Aer Turas	Eire
Racal	Racal Avionics	UK
Rafair	Royal Air Force	UK Mil
Rainbow	Queen's Flights	UK Mil
Reach	USAF Air Mobility Command	US Mil
Regal	Crown Air	Canada
Rescue	RAF Rescue	UK Mil
Richair	Rich International	USA
Rogav	Rogers Aviation	UK
Rushton	Flight Refuelling Limited	UK
Sabena	Sabena	Belgium
Sam	USAF Special Air Mission	US Mil
Saudia	Saudia	Saudi Arabia
Scandanavian	Scandanavian Airlines System	Sweden
Scanwings	Maimo Aviation	Sweden
Seychelles	Air Seychelles	Seychelles
Shamrock	Air Lingus	Ireland
Short	Short Brothers	UK
Sierra India	Arab Wings	Jordan
Singapore	Singapore Airlines	Singapore
Sky Express	Salair	Sweden
Somalair	Somali Airlines	Somalia
SouthernAir	Southern Air Transport	USA
Spantax	Spantax	Spain
Special	Metropolitan Police Air Unit	UK
Speedbird	British Airways	UK
Speedfox	Jetair Aps	Denmark
Speedpack	International Parcel Express	USA
Springbok	South African Airways	S. Africa
Starjet	Novair	UK
Stellair	Stellair	France

Callsign	Operator	Country
Sterling	Sterling Airways	Denmark
Sudanair	Sudan Airways	Sudan
Swedair	Swedair	Sweden
Swedeline	Linjeflyg	Sweden
Swedic	Swedish Air Force	Sweden
Swissair	Swissair	Switzerland
Syrianair	Syrian Arab Airlines	Syria
Tarnish	British Aerospace (Warton)	UK
Tarom	Tarom	Romania
Teastar	Trans European	UK
Tee Air	Tower Air	UK
Tennant	British Aerospace (Prestwick)	UK
Tester	Empire Test Pilots School	UK Mil
Thai Inter	Thai International Airlines	Thailand
Tibbet	British Aerospace (Hatfield)	UK
Tiger	Flying Tiger Line	USA
Tradewinds	Tradewinds Airways	UK
Trans Arabian	Trans Arabian Air Transport	Sudan
Trans Europe	Trans Europe Air Charter	UK
Trans-Med	Trans Mediterranean Airways	Lebanon
Transway	TEA Basle	Switzerl'd
Transworld	Trans World Airlines	USA
Tunair	Tunis-Air	Tunisia
Turkair	Turk Hava Yollari (THY)	Turkey
Tyrolean	Tyrolean Airways	Austria
Uganda	Uganda Airlines	Uganda
Uni-Air	Uni-Air	France
Unicorn	Queen's Flights	UK Mil
UTA	UTA	France
Varig	Varig	Brazil
Vectis	Pilatus Britten-Norman	UK
Viasa	Viasa	Venezuela
Victor Kilo	Airbus Industrie	France
Victor Yankee	Air Belgium	Belgium
Viking	Scanair	Sweden
Virgin	Virgin Atlantic	UK
Wardair	Wardair	Canada
Watchdog	Ministry of Ag. Fisheries & Food	UK Gov
West Indian	British West Indian Airways	Trinidad
White Star	Star Air	Denmark
Wigwam	CSE Aviation	UK
Woodair	Woodgate Air Services	UK
World	World Airways	USA
Worldways	Worldways Canada	Canada
Yemeni	Yemenia Airways	Yemen
Zambia	Zambia Airways	Zambia
Zap	Titan Airways	UK

Suffixes

Occasionally a suffix will be added to the group of numbers following the above callsign to denote a special characteristic of the flight. These are usually as follows although it should be noted that British Airways operate a different system.

A Added where a flight has been doubled up so that two aircraft are using the same prefix and numerals (for example one aircraft will be suffixed A and the other B).

F Aircraft carrying freight only.

P Aircraft positioning to another location with no passengers on board.

Q Aircraft details have been changed to a standard flight plan (for example the regular aircraft has been substituted with another type).

T Training flight.

X Allocate by controllers to avoid confusion when two aircraft have identical numbers even though their prefixes may differ.

'Heavy' tells controllers that the aircraft is a wide-bodied type.

Private aircraft and those operated by smaller operators such as air taxis may well use just their registration. In Britain this will consist of the prefix 'G' followed by up to four letters. On establishing first contact with a controller the pilot will give his call sign in full. For instance G-APTY would be "Golf-Alpha Papa Tango Yankee". However, the controller may well abbreviate this to "Golf-Charlie Delta" or even just "Tango Yankee".

SRA and PAR radar let-down

Occasionally, in bad visibility, a pilot may need to be 'talked-down'. SRA (Surveillance Radar Approach) may be used to give the pilot precise instructions to reach the end of the runway. Normally, the airfield has a special frequency for this and once the pilot has established contact with the controller there is a point when the controller tells the pilot not to acknowledge further instructions. From then on, the controller gives the pilot a running commentary on the aircraft's position in relation to an imaginary line drawn outwards from the runway, known as the 'centre-line'. Compass headings may be given to the pilot, to steer the aircraft, in order to get on to the centre-line. Other information given tells the pilot how far the aircraft is from the runway and what height it should be at. At a point about half a mile from the runway, the controller announces that the approach is complete. If the pilot cannot see the runway at this stage the approach must be abandoned for another, or the aircraft is diverted to another airfield. Failing to touch-down results in a 'go-around' (formerly an 'overshoot'). PAR (Precision Radar Approach) is similar, but also tells the pilot altitude and whether or not the aircraft is on the 'glide slope'.

Startup

The procedure at the start of a flight varies from airfield to airfield. On smaller airfields the pilot may start the aircraft and then ask for take-off instructions. The ground controller transmits details of which runway is in use, QFE and wind, then instructs the pilot to start taxiing to a holding point just before the end of the runway. Once the runway is clear, the controller allows the aircraft to take off and relays other instructions such as which height to climb to and when the aircraft can start turning on course. At bigger airfields, particularly those in busy flight areas, the procedure may be far more complicated and will depend to some extent on whether the flight is VFR (visual flight rules) or IFR (instrument flight rules). The first, VFR, is where an aircraft flies solely by dead reckoning. In other words, the pilot navigates by using a compass and a map, looking out of the aircraft windows for landmarks. The second, IFR, is where the pilot uses radionavigation and instruments to cover the route. Most commercial flights are IFR and such flights are always along designated airways routes. Prior to a flight the pilot files a flight plan with Air Traffic Control which is electronically distributed to controllers on the aircraft's route, who are then aware of the type of aircraft, altitude requested and destination. At commencement of the flight the pilot informs the tower that all is ready. At this stage the controller may well only say the aircraft is clear to start up and, perhaps, will give the temperature. Once the pilot informs the controller the aircraft is ready for take-off, taxiing instructions, the QFE, the runway in use and the wind details are all given. At this stage or shortly after clearance is given to the pilot, detailing destination, the airways to use and altitudes. In the UK airways are identified by colours (red, blue, green, white, and amber) with a number. After the actual take-off, the aircraft may be handed over to another controller, such as approach, before the pilot is finally told to contact 'airways' or 'information' services.

Airways

Busy air routes, such as those over Europe, are divided up into countries and regions that have central control points for all the air routes in the sectors. Although VFR flights at low altitudes can, by and large, choose the course they fly, this is not the case at higher altitudes in the airways. Now we are in the realm of 'controlled airspace' and pilots must fly along a certain course at a certain height. The airways are marked at regular points, and where they cross, by beacons on the ground. These are used for navigation purposes and also form what are known as 'compulsory reporting points': as the aircraft passes over a beacon the pilot must report to the sector controller, giving the name of the beacon, flight level (at higher altitudes the height is abbreviated, e.g., 10,000 feet becomes 'flight level one-zero-zero'), and 'forward estimate' time for the next reporting point. These points are referred to by the name of the place where the beacon is sited. Towards the end of the journey the pilot may be given a fairly complex set of instructions. Landing at major airports like London Heathrow may, at busy times, involve joining the 'stack': an imaginary spiral staircase in the sky. The aircraft joins at the top and flies a racetrack shaped circuit, slowly dropping to different flight levels until, at the bottom, it is routed to the airfield.

Flight information

Everything above 25,000 feet is 'controlled airspace' (in some regions airspace below that altitude is controlled, too). Pilots can obtain details of traffic movements in the region from the UIR (Upper-flight Information Region) service. Below that level, information is provided by the FIR (flight information region) service. Note that both the UIR and FIR are advisory services: they provide information for pilots but do not control the movement of aircraft. A further advisory service is available for small aircraft on VFR flights: LARS (Lower Airspace Radar Service). The facility is provided by the various MATZ (Military Aerodrome Traffic Zones) up and down the country. Again, they do not control flights but merely offer information regarding other aircraft in the area.

Company frequencies

Most airlines use 'company frequencies'. Any of these frequencies, however, might be used by several airline operators to contact company ground stations. An example of the use of company frequencies could be when an aircraft wishes to contact the operations department of the company base at the destination airfield, in order to give the estimated time of arrival or request special services such as wheelchairs for invalid passengers. Other messages may concern servicing required on the aircraft: instruments may need adjusting, or there may be minor technical problems that engineers will need to correct before the aircraft takes-off again.

Glossary

The following list of abbreviations and expressions are regularly used during typical transmissions between ground and air.

Abort   Abandon (i.e., abandon take-off).
AFIS   Airfield flight information service.
AIREP   Report for position and weather in flight.
Airway   Defined flight path.

AMSL	Above mean sea level.
APU	Auxiliary power unit (backup when engines are off).
ASDA	Runway accelerated stop distance.
ASI	Air speed indicator.
ATA	Actual time of arrival.
ATC	Air traffic control.
ATIS	Automatic terminal information service.
Avgas	Aviation grade petrol.
Avionics	Aircraft electronics.
Backtrack	axi back down the runway.
Beacon	Station transmitting continuous navigation signal.
CAT	Clear air turbulence.
CBs	Cumulo nimbus (thunder clouds).
Conflicting	Conflicting traffic, etc, possible collision course.
Decimal	Decimal point as in frequency, eg, 128.65MHz.
Cav-OK	Ceiling and visibility are good.
DF	Direction finding by radio.
DME	Distance measuring equipment.
Drift	Lateral movement off desired track.
ETA	Estimated time of arrival.
FIR	Flight information region.
Flameout	Total power loss on jet or turbo prop engine.
Gear	Undercarriage.
Glidepath	Line of descent on landing.
GMC	Ground movement controller.
GMT	Greenwich mean time.
Go around	Overshoot runway and re-join circuit.
GPU	Ground power unit.
Greens	Landing gear down and locked indicators.
Homer	Homing beacon.
IAS	Indicated air speed.
IFR	Instrument flight rules.
ILS	Instrument landing system.
IMC	Instrument meteorological conditions.
JET	A1 Jet and turbo-prop fuel (kerosene).
Knots	Nautical miles per hour.
LARS	Lower airspace radar service.
Localiser	Glidepath beacon.
Mach	Speed in relation to the speed of sound.
MATZ	Military aerodrome traffic zone.
METAR	Meterological report (not a forecast).
Navaid	Navigational aid.

NavCom	Combined communication and navigation radio.
Navex	Navigation exercise (training flight).
NDB	Non-directional beacon.
NOTAM	Notice to airmen.
Okta	An eighth. Used to denote cloud density.
Ops	Operations.
Orbit	Fly in a circle.
Overshoot	No longer used, see 'go around'.
Pax	Passengers (eg, 64 pax on board).
PAR	Precision approach radar.
PPO	Prior permission only (restricted airfields).
QDM	Magnetic heading.
QFE	Barometric pressure at aerodrome.
QNH	Barometric pressure at sea level.
Roll-out	Stopping distance after touchdown.
RSR	Route surveillance radar.
RVR	Runway visual range.
SAR	Search and rescue.
SELCAL	Selective calling system (activates radio by code).
SID	Standard instrument departure.
SIG	Significant.
SitRep	Situation report.
Squawk	Switch transponder on.
Squawk	Select 'identification' mode on transponder.
SRE	Surveillance radar element.
STOL	Short take-off and landing.
Stratus	Low misty cloud (often obscures runway approach).
TAI	True air speed indicator.
TACAN	actical air navigator.
TAF	Terminal area forecast.
TAR	Terminal area radar.
TAS	True air speed.
TMA	Terminal control area.
Traffic	Aircraft in flight.
UIR	Upper flight information region.
US	Unserviceable.
UTC	Universal time constant (GMT).
VASI	Runway lights angled to give a visual glide slope.
VFR	Visual flight rules.
VMC	Visual meteorological conditions.
VOLMET	Continuous weather forecast.
VOR	VHF omni-direction range beacon.

VSI	Vertical speed indicator (rate of climb).
VTOL	Vertical take-off and landing.
WX	Weather.

International Space Station

Due to its low-earth orbit, the International Space station can be heard with just a handheld scanner used outdoors with its set-top helical. It orbits the Earth about every 90 minutes, and a satellite-tracking program with up-to-date Keplers loaded will tell you exactly when it's within range of your location. The occupants also use amateur radio in the 2m amateur band on 144.800 MHz FM as recreation, usually at weekends and during their 'rest periods'. There is, at the time of writing, an amateur radio FM repeater is operational on the Space Station, with an input (uplink) frequency of 437.800MHz and an output (downlink) frequency of 145.800MHz, and this is switching into operation whenever the occupants are not using the amateur radio station themselves. So you'll typically hear ground-based stations having short contacts with each other, usually just an exchange of location and signal report, through this repeater as the station passes over you.

NASA Shuttles

Three UHF frequencies have been used by the NASA shuttles over the years but it should be noted that on some missions, the communications have been restricted to frequencies which are outside the range of ordinary scanners. However, UHF communications are heard on some flights, particularly those involving spacewalks (EVA & Extra Vehicular Activity). It must be stressed that Shuttle communications are unlikely to be heard on a scanner which is simply being used with an ordinary discone antenna. Crossed dipoles, or preferably a crossed yagi with its elements phased for circular polarisation, designed for the frequency together with a masthead pre-amplifier are advisable for best results. The following abbreviations are ones commonly used during the lift-off phase (frequently re-broadcast via the media) and during some stages of the flight.

AFSCN	Air Force Satellite Control Network.
ALT	Approach for Landing Test programme.
AMU	Astronaut manouvering unit.
APS	Alternate Payload Specialist.
APU	Auxiliary Power Unit.
ASE	Airborne Support Equipment.
ATE	Automatic Test Equipment.
ATO	Abort to Orbit.
BFC	Backup Flight Control.

CAPCOM	Capsule Communicator.
CCAFS	Cape Canaveral Air Force Station.
CCMS	Checkout, Control and Monitor Sub-systems.
CDR	Commander.
CDMS	Command and Data Management Systems Officer.
CDS	Central Data Systems.
CIC	Crew Interface Coordinator.
CIE	Communications Interface Equipment.
CTS	Call to Stations.
DCC	Data Computation Complex.
DCS	Display Control System.
DIG	Digital Image Generation.
DFI	Development Flight Instrumentation.
DFRF	Dryden Flight Research Facility.
DMC	Data Management Coordinator.
DOD	Department of Defence.
DPS	Data Processing System.
EAFB	Edwards Air Force Base.
ECLSS	Environmental Control and Life Support System.
EMU	Extra Vehicular Mobility Unit.
ESMC	Eastern Space and Missile center.
ET	External Tank.
EVA	Extra Vehicular Activity.
FAO	Flight Activities Officer.
F/C	Flight Controller.
FD	Flight Director.
FDO	Flight Dynamics Officer.
FOD	Flight Operations Directorate.
FOE	Flight Operations Engineer.
FOSO	Flight Operations Scheduling Officer.
FR	Firing Room.
FRC	Flight Control Room.
FRCS	Forward Reaction Control System.
FRF	Flight Readiness Firing.
FRR	Flight Readiness Review.
GAS	Getaway Special.
GC	Ground Control.
GDO	Guidance Officer.
GLS	Ground Launch Sequencer.
GN	Ground Network.
GNC	Guidance, Navigation and Control Systems Engineer.
GPC	General Purpose Computer.

GSE	Ground Support Equipment.
GSFC	Goddard Space Flight Center.
IG	Inertial Guidance.
ILS	Instrument Landing System.
IMF	In-Flight Maintenance.
INCO	Instrumentation and Communications Officer.
IUS	Inertial Upper Stage.
IVA	Intra Vehicular Activity.
JSC	Johnson Space Center.
KSC	Kennedy Space Center.
LC	Launch Complex
LCC	Launch Control Center.
LCS	Launch Control System.
LOX	Liquid Oxygen.
LPS	Launch Processing System.
MCC	Mission Control center.
MD	Mission Director.
ME	Main Engine.
MECO	Main Engine Cut-Off.
MET	Mission Elapsed Time.
MLS	Microwave Landing System.
MOD	Mission Operations Directorate.
MOP	Mission Operations Plan.
MPS	Main Propulsion System.
MS	Mission Specialist.
MSCI	Mission Scientist.
MSFC	Marshall Space Flight center.
NASCOM	NASA Communications Network.
NOCC	Network Operations Control center.
NSRS	NASA Safety Reporting System.
OAA	Orbiter Access Arm.
OC	Operations Coordinator.
OFI	Operational Flight Instrumentation.
OMS	Orbiter Manouvering System.
PDRS	Payload Deployment and Retrieval System.
PLT	Pilot
POD	Payload Operations Director.
PS	Payload Specialist.
RMS	Remote Manipulator System.
RTLS	Return to Launch Site.
SIP	Standard Interface Panel.
SLF	Shuttle Landing Facility.

Air ambulance helicopter; many are operated on a charitable basis.

SN	Space Network.
SPOC	Shuttle Portable On-board Computer.
SRB	Solid Rocket Booster.
SRM	Solid Rocket Motor.
SSC	Stennis Space center.
SSCP	Small Self-Contained Payload.
SSP	Standard Switch Panel.
SSME	Space Shuttle Main Engines.
TACAN	actical Air Navigation.
TAL	Trans-Atlantic Abort Landing.
TDRS	Tracking Data and Relay Satellite.
WSMC	Western Space and Missile center.

Chapter 8
Satellites on your Scanner

Many people are often amazed when they find they can hear signals from space on their scanner, whether these are voice transmissions from astronauts from the International Space Station or Space Shuttle or weather fax transmissions which a PC running suitable software can instantly convert into picture format.

Above us at this very moment, a network of satellites are beaming signals to us, pictures of our country, our continent, and even the entire globe, as seen from their 'eye' in outer space. Some of these satellites orbit the Earth a few hundred miles above us, sending us 'close-up' images. Others stay around 23,000 miles above us in geostationary orbit, transmitting

Received weather satellite picture using Timestep weather satellite program.

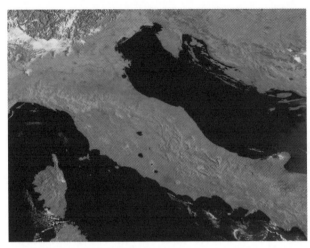

Another received weather satellite picture using Timestop weather satellite program.

images of their entire visible 'disc' as well as various sectionalised areas. Try setting your scanner to search across 137-138MHz one afternoon with your antenna having a clear 'view' of the sky above. Even with just a set-top whip, you'll soon hear the mysterious sounding bleeps from one of the many low-earth orbiting satellites as they pass otherwise silently above you.

Another 'satellite' is the International Space Station, and you'll commonly hear it on 145.800MHz when it's above your horizon. At the time of writing it carries an amateur radio repeater with the transmit frequency also on 145.800MHz, although the on-board astronauts can and often do take to the microphone themselves to chat with other radio amateurs and school stations down on the ground.

Starting out

So what do you need to decode these bleeps, and transform them into visual images? If you have a computer as well as your scanner you already have most of what you need. There are several software packages available for PCs, which normally operate by simply connecting your receiver's audio output to the input of your PC's sound card.

Many beginners start by receiving the 'low earth' orbiting satellites. You can, if you wish, simply take 'pot luck' and see what you can receive at any time, using a simple fixed antenna system such as crossed dipoles. You can alternatively run a satellite pass prediction program on your computer to tell you exactly which satellite is coming over and when.

Receivers

Although many low-cost scanners can give you a 'start', weather satellites use higher deviation than 'normal' terrestrial two-way radio signals. Your receiver needs to be able to cope with a wider than normal FM deviation,

This PC based program, JVCOMM32 uses your PC's sound card as the interface for weather satellite decoding.

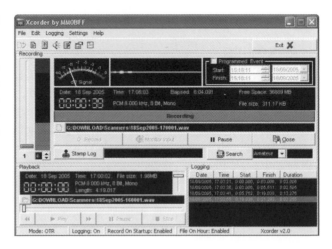

plus up to +/-3kHz of 'Doppler Shift' (the effect of increasing and decreasing frequency with movement towards and away from you) in the case of the orbiting satellites. A bandwidth of around 50kHz is ideal, but if you have a sensitive receiver then the 'Wide' FM mode on your scanner may give acceptable results, although this is often rather too wide. Some receivers, like the Icom PCR-1500, have 50kHz bandwidth selectable, and the 'best' way is to make sure your scanner has a suitable IF bandwidth, a number of scanners for both the 137MHz and 1691MHz ranges have this facility.

One limitation can be breakthrough of paging signal interference from high power transmitters operating in the adjacent VHF paging band, and some low-cost handhelds often get 'blocked' by breakthrough from these due to limitations in the scanner's performance, especially if WFM is selected. The moral here is to choose your receiver carefully.

Antennas

For reception of the orbiting 137MHz satellites, a vertical whip or discone will work but it won't give you the best results. A pair of crossed dipoles, phased for circular polarisation, are ideal, and you can improve the gain by using a pair of reflector elements beneath these, or of course by adding a masthead preamp if you don't suffer from other strong signals. For reception of the geostationary satellites such as Meteosat over Europe (at 0 deg Longitude) or the GOES series on the Western Hemisphere, a higher gain antenna is needed. A long yagi is a popular choice, in the UK this needs to be aimed due south at an elevation of around 30 degrees. Alternatively, a 1m dish can be used, fitted with either a 'probe' feed, commonly home made inside a large coffee tin, or a simple dipole element at the focal point. Larger dishes give greater gain of course, and are suitable for reception of the high quality 'Primary User Data' transmitted (see later). An antenna-mounted preamplifier is almost essential to overcome coax feeder losses at these frequencies.

Decoding

Both 'visible' and 'infra red' images are transmitted by the satellites as well as other types, but don't confuse 'weather fax' decoding (from HF signals) with 'weather satellite' decoding (from VHF/UHF signals), the two are different. Many PC-based programs also allow you to artificially 'colour' the images, to replicate blue sea and green land for example.

For orbiting satellite pass prediction, again plenty of software is available, including some as 'shareware'. For accuracy these require you to enter up-to-date Keplerian elements to allow for minor changes in the satellites' orbits, these are freely available from various sources such as the NASA site on the Internet. However many PC programs can handle this automatically, as well as automatically recording and building up a 'history' of received images. Orbiting and geostationary weather satellites transmit using APT, 'Automatic Picture Transmission', which varies the amplitude of a 2400Hz tone which is then transmitted using FM. Hence you can use the same decoder for either orbiting or geostationary satellite decoding, although most software has a number of 'pre-set modes' to cater for the slight differences in the format of these.

HRPT, High-Resolution Picture Transmission, and PDUS, Primary Data User System, formats are also transmitted in digital form. These were once limited to professional users, but 'end user' systems for the reception of these are readily available for keen devotees who'd like to receive high resolution images from the satellites.

The Meteosat images you receive on APT are processed images from a ground station, with country contours added to help identification. Images of the whole of Europe are transmitted every half hour, plus four visible close-ups of Europe every hour during daylight hours, the entire visible 'disc' sixteen times a day, plus other sections at intervals including infra-red re-transmissions taken from the GOES satellite over the USA.

Speech communications

There are a great number of other man-made satellites orbiting the Earth, some of these stay permanently in space (communication, weather and navigation vehicles); others, such as the space shuttle only stay up for a pre-determined period. Although some are automatically controlled, many are manned with crews, and communications can be readily received. For the scanner user, not all forms of transmission can be received from these space vehicles as many of the frequencies used are in the SHF band (3-30GHz). However some VHF and UHF frequencies are used and those likely to be of interest to scanner users are some of the voice communications and amateur communications relays.

Data transmissions

Many of the 'space' allocations shown do not contain voice transmissions. Many satellites transmit streams of data from on-board sensors, used for a variety of scientific measurements, and without suitable decoding equipment these signals are meaningless. The same is true of satellites used for navigation purposes. But there are many amateur satellites that transmit data where readily available decoding software is available.

Orbiting telephone satellites

With the signals from low-Earth satellites having the capability of being quite strong, a natural use for these is for personal telephone use in areas not served by a land-based cellular system. The 'Iridium' and 'Thuraya' systems ore just a couple of a number of Earth-orbiting systems for global telephone service. Each uses a number of satellites forming a cross-linked grid orbiting above the Earth. They're close enough to receive the signals of a handheld device, and act like cellular 'base stations in the sky' - where wireless signals can move overhead instead of through ground-based cells. The low earth orbit (LEO) allows ground-based phones on these systems to be much smaller than those needed for other satellites - small enough to fit in your top pocket. The satellites also keep track of the users' telephone location anywhere on the globe. The call is then relayed from satellite to satellite, until it reaches its destination: either through a local Iridium gateway and the public switched telephone network, or directly to a receiving satellite phone.

Frequencies

A band of frequencies in L-Band, between 1616.0 and 1626.5 MHz, is used as the link between the satellite and handsets. The Ka-Band (19.4-19.6 GHz for downlinks and 29.1-29.3 GHz for uplinks) serves as the link between the satellite and the gateways and earth terminals. The system uses a combination of Frequency Division Multiple Access and Time Division Multiple Access (FDMA/TDMA) signals, so although you may be able to receive the digital signals on your scanner, you won't be able to demodulate it using just FM or AM on your scanner, you'll need a sophisticated digital decoder.

Amateur satellites

Amateur satellite transmissions are among some of the easiest to receive as the satellites are designed to transmit on frequencies that are easily picked up by unsophisticated equipment. Bear in mind that the low-earth orbit satellites, like non-geostationary weather, can only be received for a few minutes at a time as they pass overhead. Occasionally their orbits take them well away from the UK and, at such times, it might not be possible to receive transmissions at all. If possible, use a set of crossed dipoles, or if you're really keen then a Yagi antenna that's automatically steerable with a rotator in azimuth and elevation. The way communications take place using these communications satellites is that an amateur transmits up to the satellite on an 'uplink' frequency; the satellite then re-transmits the signal back on a different, 'downlink', frequency. In this way it is quite easy to span large distances using VHF and UHF: communications between Europe and the Americas are quite normal on the elliptic-orbiting satellites, and throughout Europe on the low-Earth orbiting ones. This method of re-transmitting the signal is known as 'transponding', which can be either analogue or digital.

Fading signals?

As the satellite moves, its antenna effectively changes polarity in relation to the antenna of the ground station, causing the received signal to, apparently, fade away and then come back again, every few minutes. This can be overcome by using a crossed dipole antenna. Movement of the satellite also causes an effect, known as Doppler shift, which slightly alters the received frequency of the radio signal, higher in frequency as it approaches you and lower in frequency as it goes away. This means a scanner must usually be tuned a few kHz above and then a few kHz below the centre frequency to track the shifting signal. On some scanners, with small frequency step controls, this presents no great problem as, fortunately, the manual tuning control can be used to track the signal, but the inclusion of AFC (Automatic Frequency Control) on suitably-equipped scanners can be very helpful.

Amateur satellites

28.1400	AMSAT-OSCAR 51	Digital Uplink	PSK
145.2000	ARISS	Crew Contact Uplink	FM
145.8000	ARISS	Packet Downlink	AFSK
145.8000	ARISS	Crew Contact Downlink	FM
145.8500	Saudi-OSCAR 50	FM Voice Repeater Uplink	FM
145.8500	AMRAD-OSCAR 27	FM Voice Repeater Uplink	FM
145.8500	Gurwin-OSCAR 32	PacSat BBS Uplink	FSK
145.8500 - 145.8800	KiwiSat	Linear Transponder Downlink	SSB/CW
145.8600	AMSAT-OSCAR 51	PacSat BBS Uplink	AFSK
145.8600	VUSat-OSCAR 52	Beacon Uplink	CW
145.8650	KiwiSat	FM Voice Repeater Downlink	FM
145.8700 - 145.9300	VUSat-OSCAR 52	Linear Transponder Downlink	SSB/CW
145.8800	AMSAT-OSCAR 51	FM Voice Repeater Uplink	FM
145.8900	Gurwin-OSCAR 32	PacSat BBS Uplink	FSK
145.9000 - 146.0000	Fuji-OSCAR 29	Linear Transponder Uplink	SSB/CW
145.9200	AMSAT-OSCAR 51	FM Voice Repeater Uplink	FM
145.9800	MEROPE	TLM Beacon Downlink	FM
145.9900	ARISS	Packet Uplink	AFSK
429.9500	POSAT-OSCAR 28	TLM Beacon Downlink	FM

435.1500	AMSAT-OSCAR 51	FM Voice Repeater Downlink	FM
435.1500	AMSAT-OSCAR 51	PacSat BBS Downlink	AFSK
435.2200 - 435.2800	VUSat-OSCAR 52	Linear Transponder Uplink	SSB/CW
435.2250	Gurwin-OSCAR 32	PacSat BBS Downlink	FSK
435.2450	CAPE-1	TLM Beacon Downlink	FM
435.2450	KiwiSat	FM Voice Repeater Downlink	FM
435.2600 - 435.2300	KiwiSat	Linear Transponder Uplink	SSB/CW
435.3000	AMSAT-OSCAR 51	FM Voice Repeater Downlink	FM
435.3000	AMSAT-OSCAR 51	Digital Downlink	PSK
435.3520	RS-22	TLM Beacon Downlink	FM
435.7950	Fuji-OSCAR 29	Beacon Downlink	FM
435.8000 - 435.9000	Fuji-OSCAR 29	Linear Transponder Downlink	SSB/CW
436.7950	Saudi-OSCAR 50	FM Voice Repeater Downlink	FM
436.7950	AMRAD-OSCAR 27	FM Voice Repeater Downlink	FM
436.8375	CubeSat-OSCAR 55	TLM Beacon Downlink	FM
436.8450	CP3	TLM Beacon Downlink	FM
436.8475	CubeSat-OSCAR 57	Beacon Downlink	FM
436.8700	RINCON	TLM Beacon Downlink	FM
436.8700	SACRED	Linear Transponder Downlink	FM
437.1250	LUSAT-OSCAR 19	TLM Beacon Downlink	FM
437.3050	ICE Cube 1	TLM Beacon Downlink	FM
437.3250	CP2	TLM Beacon Downlink	FM
437.3450	CubeSat-OSCAR 58	Telemetry Downlink	FM
437.3850	CubeSat-OSCAR 56	TLM Beacon Downlink	CW
437.3850	KUTEsat	TLM Beacon Downlink	FM
437.4000	CubeSat-OSCAR 55	Telemetry Downlink	FM
437.4050	Mea Huaka	TLM Beacon Downlink	FM
437.4050	LIBERTAD-1	APRS Downlink	AFSK
437.4250	ICE Cube 2	TLM Beacon Downlink	FM
437.4650	CubeSat-OSCAR 58	Beacon Downlink	FM
437.4650	HAUSAT1	TLM Beacon Downlink	FM
437.4850	SEEDS	TLM Beacon Downlink	FM
437.4900	CubeSat-OSCAR 57	Telemetry Downlink	FM
437.5050	CubeSat-OSCAR 56	Packet Downlink	GMSK
437.5050	ION	TLM Beacon Downlink	FM
1268.5000	CubeSat-OSCAR 56	Packet Uplink	GMSK
1268.7000	AMSAT-OSCAR 51	FM Voice Repeater Uplink	FM
1268.7000	AMSAT-OSCAR 51	PacSat BBS Uplink	AFSK
1268.8650	KiwiSat	FM Voice Repeater Uplink	FM
1268.8800 - 1268.8500	KiwiSat	Linear Transponder Uplink	SSB/CW
2401.2000	AMSAT-OSCAR 51	FM Voice Repeater Downlink	FM
2401.2000	AMSAT-OSCAR 51	PacSat BBS Downlink	AFSK

Weather Satellite Frequencies Polar Orbiting APT (Analogue)

Satellite	Freq. (MHz)
NOAA 12	137.500
NOAA 14	137.620 (currently not operational)
NOAA 15	137.500
NOAA 17	137.620
NOAA 18	137.9125

Weather Satellite Frequencies Polar Orbiting Digital

Satellite	Freq. (MHz)	Antenna	Mode
NOAA 12	1698.0	RHCP	HRPT
NOAA 14	1707.0	RHCP	HRPT
NOAA 15	1702.5	Omni	HRPT
NOAA 16	1702.5	LHCP	HRPT
NOAA 17	1707.0	RHCP	HRPT
NOAA 18	1698.0	RHCP	HRPT
Fen Yun 1C	1700.4	RHCP	CHRPT
Feng Yun 1D	1700.4	RHCP	CHRPT

177

Weather Satellite Frequencies Geostationary;

Satellite	Position	Freq. (MHz)	Mode
Meteosat 7	0°	1694.5	HRI (digital)
		1691.0 and 1694.5	Wefax (analogue)
Meteosat 8	3° W	1691.0	HRIT (digital)
			LRIT (digital)
Meteosat 6	10°E	1691.0	HRI (digital)
			Occasional wefax

Other satellites

Several bands are allocated for space and satellite operation. Many of these bands have little, if any, activity and in recent years as technology has progressed, space communications have tended to move to higher frequencies - usually of several thousand Megahertz. Even so, there is occasionally voice traffic in some of the VHF bands. Do note, though, that they might not always carry transmissions. For instance two frequencies are shown for the NASA space shuttle but on any one flight this band might never be used. Keen space communications fans know that this side of the hobby often means much patience. If at first you hear nothing, try, try and try again. In the two very narrow navigation satellite bands it will occasionally be possible to hear either the Russian 'CICADA' system or the USA's 'TRANSIT' service. In the UK reception of 'CICADA' signal transmissions is usually possible several times a day. They are AM signals sounding like fast Morse code.

Military satcoms

Most military communications satellites are geostationary and are positioned around 35,000 Kilometres out from the equator. Their orbit is synchronised with the earth's rotation so from the ground they always appear to be in the same position. Their relatively low power output and the distance means their signals are fairly weak by the time they reach us. Ideally a small dipole or ground plane antenna cut for about 270MHz will probably give better results than a discone (a masthead preamplifier does wonders but most casual listeners probably do not want to go to those lengths). If you want to make-up a small ground plane antenna then the active element should be about 260mm long. If you use a steerable log periodic then you should get quite good results. Typical of the sort of satellites that can be heard are the American FleetSatcom series which provide global communications for the US navy. These satellites are transponders, which means that their output is on a frequency related to the input frequency (which we are unlikely to hear anyway). Most use FM, although SSB does sometimes appear. Other modes include data, radioteletype and FAX. Four of these satellites are positioned around the globe at 25 degrees and 100 degrees West, and 75 degrees and 172 degrees East (all level with the equator). In Europe, Fleetsatcom West at 23 degrees longitude can be heard, and I have heard claims that, with a suitable antenna it is also possible to hear the Indian Ocean (75 degrees East) satellite as well. The FM used has wider deviation than you will find on normal commercial bands, typically it seems to be around 25kHz. For best reception you may need to switch your scanner to the WFM mode, unfortunately most scanners are not as sensitive in this mode.

What will you hear?

The easiest transmissions to tune into are the FM voice channels. These often consist of messages being passed from ships or overseas bases back to the United States. There are also phone-patch transmissions between forces and their families on some channels. The digital/data transmissions do not appear to follow the usual commercial formats and one can assume that these are secure messages of strategic importance, which have been encoded in some way. I have heard the voice channels being switched to FAX transmission and assume that these are normal group 3 FAX transmissions, although my own station is only equipped for radio style FAX which is not compatible and so I cannot be absolutely sure.

Where to tune

Military satellite transmissions will be heard between 225 and 400MHz. Possibly the easiest voice channel to hear is Channel X-Ray on Fleetsatcom West on 261.675 MHz. This channel is used extensively for 'phone patches and is one channel which can often be received on nothing more than a telescopic antenna if you have a sensitive scanner. Bands where you should find signals are listed in:-

Military Satcoms

User	Frequency
LeaSat (US Navy)	243.850-243.900
FleetSatcom (US navy)	243.960-244.100
Marisat	248.800-249.400
LeaSat (US Air Force)	249.350-249.850
Fleetsatcom (US Navy)	260.775-268.350

FM voice traffic has also been heard on the following frequencies:
244.095, 249.550, 261.500, 261.650, 261.675, 261.950, 262.050, 262.100, 262.225, 262.300, 262.475, 263.625, 269.850 & 269.950 MHz.

Chapter 9
Your PC with your Scanner

With newer and more powerful PCs becoming available each month, many readers will either be thinking of getting one, or upgrading their existing PC to a higher-specification type. What to do with the old one, or the new one for that matter? Why not use it to help you with your hobby! The often-thought 'obvious' use is to link the RS-232 or USB serial port to your receiver to remotely control frequencies, scan banks etc. This can really 'open up' what would otherwise be a simple, manually-controlled receiver for you to listen to while you're at home at your

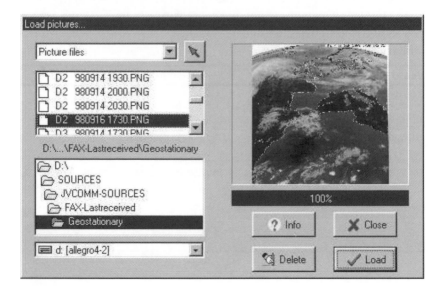

listening post. Combine this with a frequency database that you've 'imported' to your scanner memory bank control program, and you can really go to town in identifying what you hear.

Recording

What are you missing while you're away from your scanner? Are there any new active frequencies? One way is to use a 'spectrum search' to count the number of 'hits' (squelch opening occurrences) on each frequency step in a band during the day, or night. You can then go back and take a listen to what you might have been missing! But what if you've loaded up your frequency banks and search ranges, and you missed out on the action during a major event that happened while you were out at work? One way would be to use a voice-activated tape recorder, or link the squelch-operated 'tape remote' output (if your scanner has one) to the 'record' control switch input on a suitably equipped tape recorder. This way, you'll typically compress several hours' worth of monitoring, often just silence with occasional bursts of activity, into a few minutes of continuous listening.

Another way is to use a program like 'Xcorder', 'Scanner Recorder' or one of many others on your PC, where you link the audio output from your scanner to the sound card line input on your PC, the program automatically detecting received audio and recording it to your PC's hard disk for you. Add facilities such as time and date recording to each audio burst, coupled with another 'Window' on your PC controlling the scanner itself if you want to monitor more than just one frequency, and you've a rather powerful monitoring system! A little later, link the pre-recorded audio output to the audio input on your sound card while you're running a decoding program, e.g. for weather satellite decoding and display, and you've a 'time-shifted' record of what's been received in your absence. You could of course always keep your scanner linked to the PC running a recording program while you're at your listening post, to keep a record of signals. You could even be your own weatherman, with a 'movie' of otherwise still weather maps which your scanner has received every hour or so, stored and combined into a few seconds display of each. Just like you see on the TV, the difference being you did it all yourself. The possibilities are endless of course in what can, technically, be recorded, just be careful not to break the law in your scanning activities.

Pager Decoding

Many pagers in common use, apart from simple on-site systems, are alphanumeric types, where the recipient receives a text message. Even with the proliferation of cellphones, many organisations use pagers, particularly lifeboat teams, reserve fire and rescue personnel, and the like. Other users include certain personal services, including 'gorilla-gram' etc. entertainers who give their 'victims' an unexpected surprise. The vast majority of pager messages sent in the UK use a data format of POCSAG, formed by the Post Office Standards Committee Advisory Group many years ago. The other format commonly used is FLEX, which is a commercial proprietary format. Both use FM modulation, and each can be decoded using freely available software, usually using your PC's sound card as the audio interface between your scanner and the PC's data decoding capabilities.

All you need is an audio lead linking your receiver's audio output to your PC's sound card line input connector, adjusting the 'line in' level on the PC's software-controlled audio levels until you get successful decoding. You may often find that the speaker output or recorder output from your scanner will work, particularly if there's plenty of high-frequency audio content there in the first place rather than a very 'bassy' audio independent of which speaker you plug in. A much better audio output 'tap-off' is the discriminator output, if there's one present as there often is on 'up-market' base station scanners. This provides 'flat' audio output without the de-emphasis, i.e. high-frequency roll-off filtering, which normally passes via the volume control to the loudspeaker amplifier in your receiver. Otherwise you can delve in and tap off the audio yourself if you're competent in electronics, there are many freely available step-by-step scanner modification details for doing exactly this.

With most programs you can selectively filter the pager numbers to be decoded, or 'tag' a certain text string for display of messages with this included. So together with your choice of frequency, you can 'fine-tune' the system to your listening interests. For example, fire brigades use a locally-based 'pocket alerter system for calling out retained fire-fighter crews, the national frequency is 147.800MHz but I'm also told that some stations operate on 148.825 and 148.850MHz. The national paging system on 153.050MHz is also sometimes used to provide on-site fire fighters and officers with information updates. This of course could bring an added 'extra' to monitoring the going-on at occurrences around the UK. At the other end of the scale, an apprentice who in the past worked for the same organisation as myself, told me he had great fun watching the various requirements for a typical evening's 'surprise' and other activities by various local 'entertainment' firms, which were all displayed on his PC screen as the day went by!

RNLI PC Desktop Pager

Even if you don't have a pager decoder for your PC, you can still receive

lifeboat pager callouts to RNLI (Royal National Lifeboat Institution) volunteers on your PC, using a 'virtual pager'. This can be set up to start up automatically when your PC is switched on, and when active the on-screen pager beeps when an RNLI local control centre have been notified of a launch. You can choose to either all centres or just those you've selected, e.g. those in your locality. The Desktop Pager is directly linked to the RNLI's own crew pager callout system, so you'll know instantly. There are around 6,000 actual rescues each year with of course many more callouts, so you should get plenty of activity! The frequency allocations chapter in this book gives the frequencies of the actual RNLI two-way radio communications which of course will also be active after a callout. You can download the pager at http://www.rnli.org.uk/what_we_do/lifeboats/desktoppager/desktoppager

ACARS; Aircraft Control and Reporting System

This is an automatic data system used by civil airlines around the world, to transmit real-time data of flight status reports together with data and text messages involving both passengers and crew. Time of "Wheels-up", touchdown, plus ground and air speed are typical transmissions, along with text-based information. You can of course use your airband receiver linked to your PC to decode these messages, using a decoding program such as WACARS, i.e. Windows ACARS. The main frequency used in Europe is 131.725MHz, where you'll often hear periodic data bursts from a number of aircraft. Other frequencies used are 131.475MHz (Air Canada), 131.575MHz (British Midland) and 131.900 (British Airways), the latter two being mainly used for voice operational communications but with occasional ACARS.

As well as decoding messages, some programs will even plot the aircraft positions for you on an on-screen map. There are various add-on maps available to show the actual locations, as well as a useful utility program called PosFix which can automatically re-configure data for correct information and map display.

NAVTEX

Besides weather satellite decoding, detailed in the previous chapter, another interesting data form of particular use to yachtsman, is NAVTEX. This is a system used for broadcasting maritime information, ranging from shipping movements, coastguard warnings, and weather conditions. There are a number of commercial 'stand-alone' packages available for this, as well as shareware offerings. It uses a transmission system of SITOR-B, i.e. the 'broadcast' version of SITOR, Simplex Teleprinter Over Radio. SITOR-B can be decoded using a number of shareware and freeware programs. You'll need to tune your receiver to around 518kHz in LSB mode, tune back and forth a kHz or two until you get correct decode, i.e. to give the required 'beat offset' for your decoding program, and various shore-based stations transmit at different times.

Trunked Radio Tracking

If you want to 'track' MPT1327 (trunked radio) communications, you can again use your PC for this, the control data audio being fed to the sound card input of your PC. There are also freeware and shareware programs available for this, such as Trunkito', TrunkSniffer' and 'TrunkSnort'. Some will simply give you the data IDs of stations, others can also link up to your scanner's remote control connector and automatically switch channels on this for you, to track the users you're interested in. A further program, 'UniTrunker', supports control channel monitoring for Motorola Type I and II, Project 25 9600 bps, EDACS (both 4800 and 9600 bps), and MPT1327 systems.

Data decoding tips

If you use your PC for off-air data decoding, here are a few tips which I hope you'll find useful;
1) If at all possible, use the discriminator audio output from your receiver, and not the external speaker audio. Some receivers have this available as a separate output on the rear panel or facility connector. If you do use the 'normal' audio output, then also try the following;
2) Plug into the 'line in' on your soundcard and adjust the receiver volume control carefully. Alternatively, try the 'mic' input with a lower audio level from your receiver. Adjust the soundcard input level control also to suit the receiver's audio level.
3) If you're using a handheld scanner then disable the 'battery saver' (use a plug-in mains power supply to keep the receiver powered) as this

will cycle the receiver on and off, often causing the first part of a signal to be missed.

3) Keep the squelch 'open', as some scanners have a squelch opening delay which could again cause the first part of the data stream to be missed, resulting in no data at all being displayed even though the program says data is present.

4) If all else fails, read the instructions - i.e. the program's 'Help' file or other text documentation. You could find exactly what's wrong there!

PC Interference

Sometimes, PCs and radio receivers don't co-exist peacefully, due to 'noise' from the PC and its components. This is often radio-radiated hash, sometimes on the very frequencies you're trying to receive. The first thing to do is to try to keep your scanner's external antenna as far away from the PC as possible. Using a well-sited external antenna is, in any case, the single most-useful improvement you can make to your listening post. If you're using a set-top antenna on your scanner, next to the PC, you'll invariably find problems in PC noise affecting the signals you're trying to receive. You can also try temporarily unplugging your PC monitor from the PC's rear monitor output connector, and if the interference drastically reduces then try adding a couple of the commonly-available clamp-on ferrites onto the lead, one as near to the PC connector as possible and the other near to the monitor case itself. Similarly for the keyboard, and other 'peripherals' such as printers. In the past, VDU-based monitors were usually the worst culprits, LCD monitors now being far 'quieter' interference-wise. But in recent years there has been a tendency to 'miss out' the RF-filtering component on the mains input connector on the PC's power supply. Hence an external filtered connector could certainty be worth trying here.

Chapter 10
Scanner and
Accessories Review

A look at the equipment available in Britain

So, which scanner should you buy? This chapter looks at the equipment that is on offer from British dealers as well as some earlier models that are still found on the second hand market. The comments which accompany some models are those of Peter Rouse, my colleague Chris Lorek, and my own and are based either on measured technical results, on personal experience with the scanner, or general impressions gained from friends and trade sources. Although we have each owned, used, and reviewed a great many scanners the comments should not be interpreted as a recommendation for any particular model.

Buying guide

If you've read up to now, you may already have an idea of the type of scanner you'd like. Invariably, most newcomers will say they want a scanner that covers everything. But let's step back a minute, as this may not necessarily be the best for you. A 'do-everything' scanner will naturally have a large number of operating functions, together with a manual of a large number of pages – one such handheld I've used has a 100 plus page manual. An experienced scanner user will eventually get to 'grasps' with this, but a raw beginner will inevitably find it rather confusing and possibly bewildering. I'm often asked, "What's the best scanner to buy?" That's like asking a car enthusiast "What's the best car to buy?" The answer depends on what you want to do with it! There's no point in buying a Ferrari if you just want to use it once a week to go shopping. Likewise, there's no point in paying a lot of money for a scanner which many features that you won't use, and the inclusion of which could initially be off-putting. However, if you've already passed the 'introductory stage', maybe with a second hand model or one of the chap high-street type scanners with limited functions, maybe having the incorrect 5kHz rather than 12.5kHz VHF channel steps (the latter of which you probably wouldn't have bought if you you'd have read and heeded the comments on this in earlier chapters!), then please do read on.

Beginner's Scanner

Let's say you're starting out for the first time, and you're looking for a good do-everything scanner at a keen price. The answer to this is naturally very wide, because it really does depend on what your listening interests are. If they extend beyond VHF and UHF, which means HF (i.e., Short Wave) then I really would recommend an additional HF receiver, and not a 'wide range' scanner which, at least in the lower cost types, would give rather marginal performance on HF. Even something like a 'World Band' portable costing less than £100 will in my mind be money better spent than a further £100 for additional HF coverage on a scanner.

If you want continuous coverage, rather than 'banded' coverage (i.e. frequency bands with 'gaps' in between) and mode-limited scanner (i.e. AM only on the VHF airband) why not look out for a second-hand offering from AOR, Alinco, Icom or Yupiteru, the actual type being something to suit your budget and listening needs. Many specialised radio and scanner dealers (not those you'll usually find in the high street, instead look at adverts in magazines such as Radio User and Practical Wireless) which have a constantly changing second-hand stock, and buying from a dealer gives you the benefit of good advice (they want you to return!), as well as a guarantee.

It may seem an obvious statement, but there's little use in looking around for a receiver with the widest coverage possible and plenty of modes if you're only interested in say, the VHF and UHF airband ranges together with ground support frequencies ranges, and you've no interest whatsoever in HF. But then, the 1000-3000MHz range could become more and more interesting in the near future, especially if you're into advanced interests. Decide what your current interests are, and if you're just 'starting out' I'd advise against spending a fortune on the latest super-expensive receiver with it's equally super-complicated controls and buttons.

Frequency Ranges and scanning steps

Make sure you buy a receiver that suits your listening interests. For example, many budget scanners offer airband coverage on AM in their descriptions, but they offer 8.33kHz steps, nor cover the UHF military airband range. Others do cover that range, but some can't be switched from FM to AM on bands aside from 108-137MHz. If you're interested in terrestrial narrowband FM reception in the UK, most services operate on 12.5kHz steps, so ensure your scanner can accommodate this. Some scanners only allow 5kHz, 10kHz, 15kHz, or 25kHz steps, and although 5kHz steps will usually bring you 'near enough' you'll find it takes a two and a half times longer to search across a given range in the inappropriate 5kHz steps rather than in 12.5kHz increments. Some services, such as PMR446 walkie-talkies and 'eye in the sky' traffic reporter links operate with a 6.25kHz offset so this step could also be useful. If you're interested in Civil Airband you should again check that your intended scanner can handle 8.33kHz steps for VHF airband listening in Europe.

How much do you want to pay?

In general, although there are some exceptions, the more you pay for a receiver the more features you're likely to get as well as hopefully a better technical performance. If you buy a budget handheld and connect this to a rooftop antenna, particularly an amplified type, don't be surprised to find it 'falls over' with breakthrough from unwanted strong signals on other frequencies when you're trying to listen to a wanted signal. A higher price usually gets you a higher number of available memory channels, several search banks, auto-storage of active frequencies, a switchable attenuator (very useful in busy radio locations to reduce breakthrough), and maybe multi-mode reception to include SSB (Single Sideband, LSB and USB) and CW (Morse code). Few VHF/UHF commercial services use these modes, although if you're interested in listening to amateur radio DX (long distance) stations and contests, it can open up a specialised new world of listening. However these mode are essential for HF listening if you want to listen to anything other than broadcast stations on that range.

HF Coverage

Many upper-market receivers offer extended coverage to include the HF (High Frequency) range, i.e. 3-30MHz and maybe even lower frequencies, and SSB reception is essential here for utility station monitoring. But don't expect to hear much on a set-top antenna or a VHF/UHF discone on HF, you'll need a length of wire instead, preferably mounted outdoors and well in the clear. The usual adage here is "the longer the length of wire the better", but beware as many of the lower priced wideband receivers, and I'm talking of below around £300 here, won't give the performance on HF that you'll get with a similarly priced dedicated HF-only receiver. In other words, connect a good outdoor purpose-designed HF antenna, and you'll often overload the receiver with unwanted off-frequency strong signals to the detriment of what you're actually trying to listen to. If your primary interest is HF, i.e. short wave utility, amateur, and broadcast stations, then spend most of your available budget on a dedicated HF receiver and somewhat less on a further receiver to cover VHF and UHF.

Portable or Base

A handheld portable scanner is very useful in that it can double for use at home or outdoors, and will often cost less than a base receiver having similar frequency coverage. A handheld will invariably have a set-top BNC or SMA antenna connector so that you can, if you wish, use the receiver at home or out mobile with an external antenna connected in place of the set-top helical. Base scanners are, however, usually easier to operate as there's more room available on the front panel for additional buttons and controls, together with a larger frequency and channel display, and often, but not always in the case of 'budget models', give better technical performance than a handheld in terms of their ability to receive weaker signals and more importantly reject unwanted signals. Many readers use a handheld scanner for general listening, using this in the car as well as in the home. But a common limitation of using a handheld scanner in a car, when travelling at speed, is usually the low available audio output. Even if an extension speaker is plugged in, you may find there just isn't enough. But an add-on in-car low power FM transmitter can be plugged in here to allow reception through your car radio's speakers, even with front/rear fading if your system has this!

Banded or Wideband?

Many scanners are what's described as 'banded', I.e. they operate across a given number of frequency ranges rather than having a continuous, i.e. wideband, coverage. But many 'banded' scanners are quasi-wideband in disguise, for example a typical 'high-street' type scanner can be described as covering 68-88MHz, 108-137MHz, 137-144MHz, 144-148MHz, 148-174MHz, 380-450MHz, 450-470MHz and 470-512MHz, yet these bands do indeed merge on some ranges. If you're buying a set from the US, possibly whilst visiting there, the 'low band' coverage on US market scanners is usually 29-54MHz rather than 68-88MHz as found on European market 'banded' scanners. The frequency allocations elsewhere in this book give a comprehensive view on what goes on where, however if you're looking at frequency ranges provided on a typical 'high street' scanner, It does sound a lot better as a marketing statement to offer three bands between 137-174MHz, even though they are all FM with 5kHz channel spacing and that they provide continuous up/down frequency switching between them. But some gaps do exist, and I hope the brief resume below will be of use to readers considering one of these types.

68-88MHz is used by a number of services, mainly private mobile radio but also some government bodies such as the Fire and Rescue service, together with amateur radio use in the 70.0-70.5MHz range in the UK. Below 68MHz however are a number of users, such as amateurs in the 50-52MHz range, and cordless services such as low-power walkie-talkies, cordless phones and intercoms around 47MHz and 49MHz. The rest is generally military and telemetry, using modes other than those normally available on such scanner receiver systems.

87.5-108MHz is used for wideband FM broadcast radio, of minor interest in a scanner unless of course you also want to listen to this in poor audio quality on a relatively expensive receiver.

108-137MHz is used around the world for civil airband communication on AM, and is commonly called "VHF Airband". Some scanners, particularly those intended for US markets where AM is not used on other (non-aircraft) bands, have this range enabled with AM-only reception.

137-174MHz is commonly available on virtually every scanner on the market, although some only allow FM reception here. But in the UK, not all services use FM, a small number still use AM. So if your scanner doesn't have a switchable mode facility here, you'll occasionally find some rather distorted and unreadable signals in this range (as well as on the 68-88MHz range). Likewise, many cheaper scanners only allow 5kHz frequency steps in this range, whereas in the UK we use 12.5kHz and 25kHz steps and often with 6.25kHz offsets. Although in some cases you won't be able to accurately tune to the exact frequency in use, the naturally wide IF filters in these scanners often allow adequate reception. What *is* a limitation though is that the overall 'search' speed is reduced, due to the unnecessarily small frequency steps.

Just above 174MHz you'll find radio microphones, and higher still is the 'Band III' range (extending up to 225MHz). This used by a number of services, particularly public transport such as railways and buses/coaches. Part of this band is also used for DAB, Digital Audio Broadcasting.

The frequency spectrum *225-400MHz* is virtually exclusively used for military airband communications, and is commonly called "UHF airband". This also uses AM but typically with 50kHz minimum channel spacing. If you're interested in military airband monitoring, ensure this range is covered on your scanner, and that you can receive AM signals on it - many scanners with 'airband' coverage claimed only cover VHF Airband. *400-406MHz* is used for satellite communication, but *406-470MHz* is used by a very large number of services in the UK. These range from low-power walkie-talkies and alarms, to countrywide communication users such as security forces. You can usually dismiss the range from about 470MHz upwards on a scanner, as this spectrum, up to 854MHz, is currently used for TV broadcasting, and most if not all further use after the analogue TV switchover are invariably likely to be digital; modes. Above 854MHz are services such digital cellular phones.

I hope the above will help potential scanner buyers in deciding whether the extra £100 or so extra is worthwhile spending to get a wideband 'all-coverage' scanner or not

Wideband scanners

If you don't want to miss a thing, these are the types to go for, and in a number of cases you may find you're not paying a premium. The small Icom IC-R2 for example is a compact wideband model but with a price tag actually less than a number of banded scanners. What you'll normally find however is that extra facilities, such a bandscope, alpha-tagging of memory channels, and the inclusion of SSB, increases the price accordingly. If you're choosing a sophisticated model as your first set, ensure you're happy in knowing how to use it, preferably in the shop with a 'hands-on' test, before parting with your money.

Dedicated Airband scanners

Although the majority of scanners cover the civil airband, many of them offer performance that is something of a compromise. Airband transmissions are AM mode and the ideal circuitry for AM differs to that for FM. Although the majority of dual or multimode scanners offer adequate performance on airband, they rarely achieve the same level of performance as dedicated airband scanners such as the early Sony, WIN or Signal models where the RF, IF and AGC circuits are optimised for the band and mode. The IF bandwidth particularly needs to be wide enough to accommodate offset carriers which are sometimes encountered on the civil airband range.

Where to buy from

The most important advice is to buy from a dealer who will have proper facilities and stock to repair or replace any equipment he sells you. If your new scanner is going to go wrong in the first few years of use it will statistically do so in the first few weeks you own it, and there can be nothing more infuriating than waiting for weeks or even months for it to be repaired or to await a replacement. The EU Sale and Supply of Goods to Consumers Regulations will of course help here, in that you'll simply be able to get a refund and buy elsewhere. If you buy a scanner from a dealer who specialises in this kind of equipment (most amateur radio equipment shops do have facilities) then repair should be fairly fast. On the other hand if you buy from the high street shop that sells a few TV's, radios and household appliances they are unlikely to have the kind of sophisticated equipment nor the experience to repair or re-align sophisticated VHF/UHF equipment. Never mind what the salesman says, it is a specialised area of servicing. With a specialised dealer you will also get proper advice on choosing a scanner. Salesmen who work in the majority of amateur radio type shops are usually enthusiasts and know their products well. On the other hand I could fill a book with some of the fantastic claims and howlers I have heard pour from the mouths of shop assistants selling scanners in Hi-Fi/TV shops.

Buying used equipment

Used scanners are commonly available from specialised radio dealers, these often having been 'traded in' as a customer upgrades to a more sophisticated model. Buying a second hand scanner from such a dealer will usually give you the best assurance of 'getting what you want' as well as usually a 3-month guarantee. Second hand equipment is also advertised in the readers adverts sections of magazines devoted to hobby radio, as well as sometimes in local 'free' sales papers. When buying a

Secondhand scanners are readily available from dealers

used scanner from an individual rather than a dealer, if at all possible take an experienced scanner user with you. The first thing to check is the frequency coverage and, if it is important to your choice of listening, whether or not the mode is selectable on all bands. Also take a look at it's general appearance, if it has a number of scratches or other damage then it's obviously seen a rough time, however most current scanners stand up to this well and will often operate perfectly. But watch out for any evidence of internal tampering, such as chewed-up screw heads on the case. It may have had the attention of a 'screwdriver expert' inside, possibly to the detriment on the set's performance. The worst offenders are often those people with a little knowledge of radio circuitry and so be doubly cautious if the seller seems to be well versed in technology. Obviously you should ask for a demonstration of the scanner and don't be afraid to ask if the owner has had any problems with it or had it repaired.

Prices

These constantly change as far-eastern manufacturing technology changes as well as currency fluctuations. The best guide is to consult a recent copy of a dedicated radio magazine, where you'll often find dealers advertising such equipment. Some also have a couple of pages purely of dealers' lists of second hand equipment, and the readers ads will undoubtedly have several offers you might be tempted with.

The following section on actual equipment has been updated to take account of new scanners available, and the technical performance figures, where given below, are real measurements of a typical and representative sample of the scanner model in each case. All the results are from measurements made in exactly the same way, and are independent of any manufacturer's 'claims' or 'specifications', which due to the different methods of each manufacturer's specifications can often give a meaningless comparison! For the technically minded, the results given here are measured as:

Sensitivity; Input signal level in uV pd giving 12dB SINAD.

Adjacent channel selectivity; Increase in level of interfering signal, modulated with 400Hz at 1.5kHz deviation, above 12dB SINAD reference level to cause 6dB degradation in 12dB on-channel signal.

Blocking: as adjacent channel selectivity.

Intermodulation Rejection; Increase over 12dB SINAD level of two interfering signals giving identical 12dB SINAD 3rd order inter-modulation product (i.e., the unwanted on-channel signal produced from two off-channel signals). All measurements, except sensitivity where stated, were taken on 145MHz NFM to ensure uniformity (apart from airband only sets where 125MHz AM was used).

Alinco DJ-X1D

Type: Handheld
Coverage: 500kHz 1300MHz
continuous
Modes: AM, FM, Wide FM
Sensitivity: 20MHz 0.37uV AM,
                    0.20uV FM 145MHz
                    0.16uV AM, 0.13uV
                    FM 435MHz 0.18uV
                    AM, 0.12uV FM
                    934MHz 0.20MHz
                    0.20uV FM
Adjacent Channel: 12.5kHz 18.9dB 25kHz
                    40.9dB
Blocking: 100kHz 40.9dB 1MHz
                    80.6dB 10MHz 87.1db
Intermodulation: 25/50kHz 24.0dB 50/
                    100kHz 27.9dB

A wideband scanner, giving plenty of listening scope in a slightly chunky but still compact handheld sized case. Some 'European' versions of this model come supplied with coverage of the amateur bands only, all other frequency ranges being locked out unless you tap in the right numbers (which the UK distributors supply on an information sheet with the set), so beware if you're buying from abroad. The 'D' suffix of the DJ-X1D is supposed to mean it's an improved version of the DJ-X1, better suited to handling strong signals. It does give good performance from 'out of band' signals, but connect an outside antenna and you'll most likely have a lot of problems from signals up to a few channels away from the one you're tuned to.

Alinco DJ-X3

Type;	Handheld
Coverage;	100kHz - 1300MHz
Modes;	AM/FM/WFM
Channels;	700

A very compact and easy-to carry around scanner, weighing just 14.5g without batteries. With a built-in audio descrambler (inverter), stereo FM reception if you plug headphones in, and 8.33kHz steps for airband.

Alinco DJX7

Type;	Handheld
Coverage;	100kHz - 1300MHz
Modes;	AM/FM/WFM
Channels;	1000

Similar in size to the DJ-X3, again with a built-in audio descrambler (inverter), stereo FM reception if you plug headphones in, and 8.33kHz steps for airband, tone squelch, plus the facility to remotely program it with an optional cable and free downloadable software.

195

Alinco DJ-X10

Type: Handheld
Coverage: 500kHz 2000MHz
                 continuous
Modes: AM, FM, Wide USB,
                 LSB, CW
Channels;   1200
Sensitivity: 20MHz 0.13uV
                 SSB, 0.23uV
                 AM, 0.15uV
                 FM 145MHz
                 0.18uV SSB, 0.28uV
                 AM, 0.18uV FM
                 435MHz 0.32uV SSB,
                 0.28uV FM 950MHz
                 0.27uV FM
Adjacent Channel: 12.5kHz 24.9dB
                 25kHz 47.7dB
Blocking: 100kHz 61.1dB
                 1MHz 78.8dB
Intermodulation: 25/50kHz
                 61.8dB 50/100kHz
                 59.4dB

A 'go anywhere on any frequency and any mode' scanner in a slim case. It's neat, easy to use, and gives good performance on air. The scanner has both 'beginner' and 'expert' operating modes as well as auto-mode and auto-step functions, which make it useful for even the beginner. There are plenty of useful memory scanning facilities, also a facility to alpha-tag memory channels plus a band-scope display. There's an auto-store facility, but this repeatedly stores channels already found until the memory bank is filled up, not a lot of use. The inclusion of SSB is useful for HF listening, although it's naturally not up to the performance of a purpose-designed HF receiver.

Alinco DJ-X2000

Type;   Handheld
Coverage;   100kHz - 2149.99MHz
Modes;   AM/NFM/WFM/ LSB/USB/CW
Channels;   2000

With 'flashtune' to tune your receiver to the frequency of a nearby transmitter and a unique facility that is claimed to locate hidden transmitter 'bugs'. Channel scope, CTCSS identification, built in recorder for up to 160 seconds of audio, analogue inverter descrambler, and a built-in frequency counter and field strength meter,

196

AOR AR-1500EX

Type:	Handheld
Coverage:	500MHz 1300MHz continuous
Modes:	AM, FM, SSB with built-in BFO
Channels:	1000
Sensitivity:	20MHz 0.18uV 0.12uV FM 145MHz 0.22uV AM 0.13uV FM 435MHz 0.45uV AM 0.38uV FM 934MHz 0.57uV FM
Adjacent Channel:	12.5kHz 34.0dB 25kHz 53.5dB
Blocking;	100kHz 48.0dB 1MHz 75.0dB 10MHz 97.0dB
Intermodulation:	25/50kHz 45.0dB 50/100kHz 48.0dB

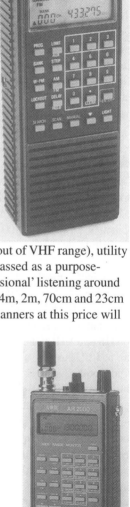

This wideband handheld scanner sells at only a little more than AOR's similar, but AM/FM only, AR-2000 scanner, yet it gains a lot from its switchable BFO (Beat Frequency Oscillator). This lets you use the set to far greater advantage on HF, for example allowing reception of HF Airband (used when craft are over the oceans, out of VHF range), utility stations, and radio amateurs. The scanner can't be classed as a purpose-designed SSB receiver, but it's certainly OK for 'occasional' listening around on this mode. You might even be able to tune into 6m, 4m, 2m, 70cm and 23cm SSB stations during contests which very few other scanners at this price will let you do.

AOR AR-2000

Type:	Handheld
Coverage:	500kHz 1300MHz continuous
Modes:	AM, FM, Wide FM
Channels:	1000

This set is essentially identical to the AR-1500 but without the BFO facility.

AOR AR-2002

Type:	Base/mobile
Coverage:	25 550MHz, 800 1300MHz
Modes:	AM, FM, Wide FM
Channels:	20

This early model from AOR looks broadly similar in styling to the superb but quite different AR-3000A. The AR-2002 is reputed to give good performances on-air, as well as being easy to use. A very wide band coverage will let you listen to plenty of things, but you only have 20 memories available to store all those interesting frequencies in. It does, however, have an RS-232 remote control connector, which allows you to use the power of your PC for this. If you're after something second hand to use at home and don't mind the limited built-in memories, you could get a good bargain.

AOR AR-2800

Type:	Base/mobile
Coverage:	500kHz 600MHz, 800MHz 1300MHz
Modes:	AM, FM, Wide FM, SSB
Channels:	1000
Sensitivity:	20MHz 0.57uV AM 145MHz 0.31uV AM, 0.20uV FM 435MHz 0.28uV FM 934MHz 0.29uV FM
Adjacent Channel:	12.5kHz 33.0dB 25kHz 33.0dB
Blocking	100kHz 56.5dB 1MHz 75.5dB 10MHz 97.0dB

Like the AR-1500 handheld, it packs in a lot of frequency coverage at a reasonable price, and has a switchable BFO so you can tune into SSB signals on HF and VHF/UHF bands. A nice 'extra' touch is a vertical bargraph made up of LEDs to give a relative display of the received signal level. The very wideband coverage, with just a 'gap' in between (which is used for TV broadcasting in the UK) should let you tune into lots of signals, the sensible number of 1000 channels being more than enough to suit most user's needs in

terms of frequency storage. Watch out though, it isn't a purpose-made HF SSB receiver, it'll 'curl up' if you put a well-sited outdoor antenna on it, although you can switch in an internal attenuator to reduce the effects of this. For searching around on VHF/UHF, either AM or FM, an 'AF scan' toggle switch is hidden at the back of the case.

AOR AR-3000A

Type:	Base/mobile
Coverage:	100kHz 2036MHz
Modes:	AM, FM, Wide FM, CW, LSB, SSB
Channels:	400
Sensitivity:	20MHz 0.12uV SSB 0.34uV AM 0.16uV FM 145MHz
	0.17uV SSB 0.48uV AM 0.24uV FM 435MHz 0.18uV SSB
	0.51uV AM 0.25uV FM 934MHz 0.26uV FM
Adjacent Channel:	12.5kHz 42.0dB 25kHz 52.5dB
Blocking	100kHz 67.0dB 1MHz 79.0dB 10MHz 94.5dB
Intermodulation:	25/50kHz 41.5dB 50/100kHz 41.0dB

If you think of what you'd like in a mobile or base scanner and then make a 'wish list', this one's probably rather a lot of them. With coverage up to 2036MHz, you could even have a go at tuning in some of the geostationary communications satellites, and at the other end of the spectrum, short wave broadcast, utility, amateur, and HF airband and marine signals. Add a PC running 'Searchlight' software available from AOR, and you have a very, very powerful listening station. It's little wonder the UK government are reputed to have bought a large number of these for themselves. It isn't cheap, but it's worth every penny, secondhand models are very sought after.

AOR AR-5000

Type:	Base
Coverage:	10kHz 2600MHz
Modes:	AM, FM, Wide FM, CW, LSB, SSB
Channels:	1000
Sensitivity:	20MHz 0.23uV SSB 0.81uV AM 0.43uV FM 145MHz
	0.17uV SSB 0.46uV AM 0.24uV FM 435MHz 0.19uV SSB
	0.26uV FM 950MHz 0.30uV FM
Adjacent Channel:	12.5kHz 50.9dB 25kHz 54.2dB
Blocking	100kHz 75.4dB 1MHz 92.6dB 10MHz 95.2dB
Intermodulation:	25/50kHz 76.0dB

About the 'ultimate' in desktop scanners, this one's got the lot. There's also an AR5000+3 model with added synchronous AM, AFC (Automatic Frequency Control) and a noise blanker. A staggeringly wide frequency range, selectable bandwidths of 3, 6, 15, 30, 110 and 220kHz and even the facility of adding an optional 500Hz or 1 kHz filter for CW or data. It gives excellent on-air performance with very good rejection of unwanted signals. A remote control port lets you couple a PC up to this superb receiver to give an extremely capable and very powerful monitoring tool.

AOR AR-8000

Type:	Handheld
Coverage:	500kHz 1900MHz continuous
Modes:	AM, FM, Wide FM, USB, LSB, CW
Channels;	1000
Sensitivity:	20MHz 0.22uV SSB, 0.33uV AM, 145MHz 0.19uV SSB, 0.28uV FM, 435MHz 0.36uV SSB, 0.32uV FM, 950MHz 0.32uV FM, 1600MHz 2.89uV FM
Adjacent Channel:	12.5kHz 38.3dB, 25kHz 47.5dB
Blocking:	100kHz 60.8dB 1MHz 68.9dB
Intermodulation:	25/50kHz 73.8dB 50/100kHz 62.6dB

Plenty of operating modes, including a 'newuser' mode for simple operation which you can switch to

'expert' as you progress. An auto-mode facility in band segments is very useful, the scanner changing mode as needed when you tune around, as well as an alphanumeric text display giving helpful info while you're programming, and a band-scope. A auto-store facility is fitted which works well, and a remote control facility. You can password protect the upper 500 memory channels to save prying eyes, although a hard microprocessor reset disables this yet retains the memories should you forget your four-digit lock code.

AOR AR-8200

Type:	Handheld
Coverage:	530kHz 2040MHz continuous
Modes:	AM, FM, Wide FM, USB, LSB, CW
Channels;	1000
Sensitivity:	20MHz 0.13uV SSB, 0.53uV AM, 145MHz 0.13uV SSB, 0.18uV FM, 435MHz 0.22uV FM, 800MHz 0.47uV FM, 1700MHz 0.61uV FM
Adjacent Channel:	12.5kHz 25.7dB, 25kHz 44.4dB
Blocking:	100kHz 65.3dB, 1MHz 71.3dB, 10MHz 86.1dB
Intermodulation:	25/50kHz 53.7dB, 50/ 100kHz 52.7dB

Packed to bursting with 'bells and whistles', this one's probably the most feature-filled handheld scanner around. A large dot-matrix LCD gives a text display for memory alpha-tagging as well as a band-scope. A useful facility of the latter is a 'peak hold' where you can leave the set scanning and return later to see what activity there's been. A vast number of scanning modes including carrier, speech, level etc., even a remote control port. A detachable medium wave bar antenna is fitted which gives good portable reception in this range. There are a number of small plug-in options available, including an add-on memory unit (4 groups of 1000 channels each), a voice inverter, a CTCSS unit, an audio tone eliminator and even a digital audio recorder. The inclusion of selectable 8.33kHz channel spacing steps makes this one future-proof for VHF airband.

AOR AR-8600

Type;	Base
Coverage;	530kHz- 2040MHz
Modes;	WFM, NFM, SFM, WAM, AM, NAM, USB, LSB, CW
Channels;	1000

A versatile and high-performance receiver with a TCXO (Temperature Controlled Crystal Oscillator) for good stability, particularly for SSB. Remotely controllable with free-to-obtain software from the AOR web site.

AOR AR-One

Type;   Base
Coverage;   0.01-3300
               Modes;
               AM, NFM,
               WFM,
               USB, LSB,
               CW, Data
Channels;   1000

A multi-everything scanner!

Bearcat BC-890XLT

(Not illustrated)
Type: Base
Coverage: 29 54, 108 512MHz, 806 1300MHz
Modes: AM, FM
Channels: 200
Sensitivity: 29MHz 0.32uV FM 145MHz 0.34uV FM, 435MHz
               0.55uV FM 934MHz 0.27uV FM
Adjacent Channel: 12.5kHz 4.8dBdB, 25kHz 56.4dB
Blocking   100kHz 68.3dB, 1MHz 79.3dB, 10MHz 86.3dB

A large sized base scanner from Bearcat, offering quite an impressive looking desktop set. It's packed with handy operating facilities such as an automatic store facility and quick access for scanning your favourite banks of frequencies, which are arranged as 20 banks of 10 channels each. 'Turbo Scan' give a fast scan rate of around 100 channels per second, meaning you shouldn't miss very much. The frequency steps can usefully be programmed to either 5kHz, 12.5kHz, or 25kHz, you're not 'stuck' with 5kHz steps on VHF like other scanners which are also designed for the US market.

Bearcat BC-2500XLT

Type: Handheld
Coverage: 25 550, 760 1300MHz
Modes: AM, FM, Wide FM
Channels: 400

This one's a 'full-feature' model from Bearcat, with 400 channels arranged into 20 banks, and an 'automatic store' to help fill these for you with locally active frequencies. 'Turbo scan' mode lets you hunt around at almost 100 channels per second, and for manual tuning a rotary knob is fitted. The set automatically switches between AM and FM depending on the sub-band selected (AM for Civil and Military airband plus 25-29MHz, FM otherwise), which could either be useful or limiting, depending on your needs.

Uniden-Bearcat
UBC30XLT

Type;   Handheld
Coverage;   87.5-107.9 WFM, 108-
               136.9875 AM, 137-
               173.99 FM
Modes;   WFM/AM/FM auto-
               selected according to
               band
Channels;   200

An economic scanner covering VHF
Airband with 8.33kHz steps, marine and
VHF PMR public service frequencies,also
includes FM broadcast.

Uniden-Bearcat UBC
68XLT

Type;   Handheld
Coverage;   66-88 137-174,
               406-512
Modes;   FM
Channels;   80

A simple and easy to use scanner, it
comes with five pre-programmed UK
search bands to help you get started
quickly. Of interest to the non-
airband listener.

Uniden-Bearcat UBC-
72XLT

Type;   Handheld
Coverage;   25.0-87.2625,
               108.0-173.99,
               406.0-512.0
Modes;   AM/FM
Channels;   100

The big advantage of this
economically priced scanner is its
'Close Call' feature – make sure this
is enabled if you're thinking of
buying one! Includes 8.33kHz steps
for VHF airband.

Uniden-Bearcat UBC92XLT

Type;	Handheld
Coverage;	25-54,108-174,406-512,806-956
Modes;	AM/FM
Channels;	200

Another economic scanner, limited to 5kHz steps on VHF FM.

Uniden-Bearcat UBC105XLT

Type;	Handheld
Coverage;	29-54, 108-174, 406-512
Modes;	AM/FM
Channels;	100

A low cost scanner, again limited to 5kHz steps on VHF FM but including 8.33kHz steps for VHF airband.

Uniden-Bearcat UBC180XLT

Type;	Handheld
Coverage;	25.0-87.2625, 108.0-173.99, 406.0-512.0, 806-960
Modes;	AM/FM
Channels;	100

This is marketed as a 'Sports' scanner, with a very fast scan rate of up to 300 channels per second, albeit with fixed non-UK tuning steps dependant upon the band section selected. 8.33kHz steps on airband.
CTCSS and DCS decode facilities are built in.

Uniden-Bearcat UBC3300XLT

Type;   Handheld
Coverage;   25.0-512, 806-1300
Modes;   AM/FM
Channels;   1000

Offering 'Trunk Tracking' (EDACS, Motorola, E.F. Johnson but not MPT1327) and usefully having no band 'gaps' apart from UHF TV.

Uniden UBC780XLT

Type;   Base
Coverage;   25-512, 806-956, 1240-1300
Mhz Modes;   AM/FM/WFM
Channels;   500

Another 'Trunk Tracking' (EDACS, Motorola, E.F Johnson but not MPT1327) radio but again usefully having few band 'gaps'.

Uniden UBC-278CLT

Type;   Base
Coverage;   0.520-1.720, 25-174, 406-512,
Mhz Modes;   AM/FM/WFM
Channels;   100

A base scanner with pre-set tuning steps (non-UK, i.e. 10kHz on medium wave and 5kHz on VHF FM), 12.5kHz steps on VHF airband.

Black Jaguar MkIV

Type: Handheld
Coverage: 28 30MHz, 60 88MHz, 115
                     178MHz, 210 260MHz, 410
                     520MHz
Modes: AM, FM
Channels: 16

An early handheld scanner, which can be picked up on the second hand market, although rather heavy and bulky by today's standards. It's included here only because it can sometimes be found on sale by private sellers, and could be useful as a second unit to keep on a few channels of interest.

205

Fairmate HP-2000

Type: Handheld
Coverage: 100kHz 1300MHz
Modes: AM, FM, Wide FM
Channels: 1000

Looking remarkably similar to the AR-2000, it also has many of the same facilities, although the lower tuning range has been extended to 100kHz.

GRE PSR-255

Type;   Handheld
Frequency Range; 26-54, 66-88, 137-174, 380-512MHz
Modes;   FM only
Memories;   50 + 1 monitor

Powered by 6 x AA batteries or 9V external power
An economic scanner, handheld size, limited in memories but could be a reasonable 'starter' or a second unit to keep in the car. Provides fixed 12.5kHz steps on UHF so it's not suitable for monitoring PMR446 etc.

GRE PSR-282

Type;   Handheld
Frequency Range; 66-88, 118-137, 137-174, 380-512MHz
Modes   AM, FM
Memories;   200

Powered by 4 x AA batteries or 6V external power
Another economic scanner but with the useful inclusion of VHF airband with 8.33kHz steps. Fixed 12.5kHz steps on UHF so it's not suitable for monitoring PMR466 etc. The scanner is usefully is supplied with two interchangeable battery compartments, one for normal batteries and one for rechargeable batteries, handy for a quick change when you're out and about.

GRE PSR-295

Type;	Handheld
Modes;	AM/FM
Frequency Range;	25 - 88, 118 - 137, 137 - 174, 216 - 225, 225 - 512, 806 960, 1240 -1300 MHz
Modes;	AM, FM
Memories;	1,000

Powered by 4 x AA batteries or 6V external power. A simple to use scanner with correct 8.33 kHz steps on VHF Civil airband and coverage of UHF Military airband as well. 12.5 kHz steps on VHF high band (137-174MHz) and UHF (400-512MHz) thus again PMR446 etc. channels aren't accommodated. The scanner is usefully is supplied with two interchangeable battery compartments, one for normal batteries and one for rechargeable batteries, handy for a quick change when you're out and about. An internal RF attenuator can usefully be programmed on a channel-by-channel basis.

Icom IC-PCR100

Type:	PC Controlled
Coverage:	500kHz 1300MHz
Modes:	AM, FM, Wide FM
Sensitivity:	20MHz 0.77uV AM 0.32uV FM 145MHz 0.55uV AM 0.25uV FM 435MHz 0.46uV AM 0.23uV FM 1300MHz 0.35uV FM
Adjacent Channel:	12.5kHz 48.6dB 25kHz 59.3dB
Blocking	100kHz 71.9dB 1MHz 82.2dB 10MHz 78.4dB
Intermodulation:	25/50kHz 47.2dB 50/ 100kHz 55.9dB

With all the PC facilities you'd expect including a virtual front panel and bandscope, auto-mode, memory tagging and so on. FM filter bandwidths of 6kHz and 15kHz are useful for 12.5kHz and 25kHz channel spacing, and a switchable WFM filter bandwidth down to 50kHz from 230khz gives potential for VHF weather satellite monitoring. Built-in CTCSS decoder with tone scan, and quite reasonable on-air performance.

Icom IC-PCR1000

Type: PC Controlled
Coverage: 500kHz 1300MHz
Modes: AM, FM, Wide FM, USB, LSB, CW

'Big brother' to the IC-PCR100, this adds SSB capabilities together with three different types of virtual front panel display with the supplied Icom software. Wit the very comprehensive scanning facilities you'd expect, although the band-scope mutes received audio while its searching around.

Icom IC-R1

Type: Handheld
Coverage: 100kHz 1300MHz
Modes: AM, FM, Wide FM
Channels: 100
Sensitivity: 30MHz 0.34uV AM
                 0.32uV FM 145MHz
                 0.31uV AM 0.25uV FM
                 435MHz 0.42uV AM
                 0.27uV FM 934MHz
                 0.38uV FM
Adjacent Channel: 12.5kHz 21.8dB 25kHz
                 33.0dB
Blocking   100kHz 31.0dB 1MHz
                 67.5dB 10MHz 79.5dB
Intermodulation: 50/100kHz 24.0dB

Icom's first handheld scanner squeezes in wideband coverage and a number of nice features like switchable steps of 0.5, 5, 8, 9, 10, 12.5, 20, 25 and 50kHz for tuning increments. Another potentially useful mode is 'Auto-Memory Write Scan' where you can set the receiver searching across a band, and it will automatically store the first 19 channels where it finds a signal. Unfortunately, the set suffers from the manufacturer's attempts of trying to squeeze an lot in, and it suffers from strong-signal overload from adjacent channels unless an internal crystal filter addition modification has been performed.

Icom IC-R2

Type: Handheld
Coverage: 500kHz 1310MHz
Modes: AM, FM, Wide FM
Channels: 400

A neat and very portable scanner, lightweight and with the footprint of a credit card, and very reasonably priced. It's powered by two AA batteries so spares are easily carried for an extended listening spell away from home. 50 extra memory channels are fitted as band-edge channels for searching, and a

CTCSS tone-scan lets you see which tone is in use on the channel you're monitoring as well as the set using tone squelch to monitor only signals with the correct tone. A cloning facility lets you upload and download channels with another IC-R2 or a PC.

Icom IC-RX7

Type;	Handheld
Frequency Range;	0.150-1300MHz
Modes;	FM, WFM, AM
Memories;	1650
Channel Steps;	5.0, 6.25, 7.5, 8.33, 9.0, 10.0, 12.5, 15, 20, 25, 30, 50, 100, 125 and 200 kHz

Powered by internal 3.7V 1100mAh rechargeable Li-Ion battery pack. A very stylish scanner and slim enough to pop into your pocket, this is a fully-featured receiver complete with a built-in ferrite antenna for medium wave reception and an earphone cord antenna for FM broadcast listening. It includes a voice squelch facility, search with automatic memory write, sub-tone and digital coded squelch with sub-tone and DCS code search, and a PC remote port for programming and cloning.

Icom IC-R10

Type:	Handheld
Coverage:	500kHz 1300MHz
Modes:	AM, FM, Wide FM, SSB, CW
Channels:	1000
Sensitivity:	20MHz 0.22uV SSB 0.36uV AM 0.23uV FM 145MHz 0.21uV SSB 0.31uV AM 0.18uV FM 435MHz 0.38uV SSB 0.34uV FM 950MHz 0.35uV FM
Adjacent Channel:	12.5kHz 21.9dB 25kHz 50.6dB
Blocking 100kHz	59.4dB 1MHz 83.9dB 10MHz 87.8dB
Intermodulation:	25/50kHz 52.6dB 50/100kHz 53.1dB

The 'big brother' to the IC-R1 and R2, this one includes SSB receive and a remote port for full remote control from a PC, as well as memory upload/

download. A unique facility of being able to listen to one channel while the set automatically searches for the next busy channel speeds listening up remarkably. An 'easy' mode gives you a rapid scan of pre-programmed ranges, there's also an auto-write scan fitted to locate and store new channels for you as well as a bandscope. Reasonable on-air performance although connecting an external antenna could bring problems in busy areas.

Icom IC-R100

Type:	Base/mobile
Coverage:	100kHz 1800MHz
Modes:	AM, FM, Wide FM
Channels:	100
Sensitivity:	30MHz 1.12uV AM 0.51uV FM 145MHz 0.39uV AM 0.18uV FM 435MHz 0.58uV AM 0.27uV FM 934MHz 0.23uV FM
Adjacent Channel:	12.5kHz 42.3dB 25kHz 62.3dB
Blocking 100kHz	75.5dB 1MHz 91.5dB 10MHz 105dB
Intermodulation:	50/100kHz 55.0dB

A quality receiver in a car-radio sized case. Useful facilities such as switchable VHF preamp and AFC (Automatic Frequency Control) helps in weak signal reception of orbiting satellites, the

1800MHz upper frequency coverage accommodates geostationary satellites as well.

Icom ICR-7100

Type:	Base
Coverage:	25MHz 2000MHz
Modes:	AM, FM, Wide FM, LSB, USB
Channels:	900

This receiver, like its predecessor the ICR-7000, has been the choice of 'serious' VHF/UHF listeners for some time, including many professional users. Its SSB reception capability lets you listen to plenty of 'extra' activity, and the 2000MHz upper limit tuning (although the specifications are only guaranteed up to 1300MHz) again puts it in the realms of the 'serious listening' category. Remote computer control facilities are also available via a CI-V port

Icom ICR-8500

Type: Base
Coverage: 100kHz 2000MHz
Modes: AM, FM, Wide FM, LSB, USB, CW
Sensitivity: 20MHz 0.14uV SSB 0.38uV AM 0.18uV FM 145MHz
                    0.17uV SSB 0.45uV AM 0.21uV FM 435MHz 0.15uV SSB
                    0.19uV FM 1800MHz 0.43uV FM
Adjacent Channel: 12.5kHz 58.8dB 25kHz 67.0dB
Blocking 100kHz   78.4dB 1MHz 94.3dB 10MHz 95.9dB
Intermodulation: 25/50kHz 75.2dB 50/100kHz 74.6dB

A worthy and equally powerful successor to the IC-R7000 and 7100 family. More of a serious receiver than a desktop scanner, this model adds HF coverage, an IF shift control and APF (automatic peak filter) being useful here. PC control facilities add to the set's already powerful scanning and monitoring capabilities.

Icom ICR-9000

Type: Base
Coverage: 100kHz 2000MHz
Modes: AM, FM, Wide FM, SSB, CW, FSK
Channels: 1000

One for the 'serious listener' and a receiver certainly also aimed at the professional user. It's a full-blooded base station VHF/UHF communications receiver more than a 'scanner', and the 75mm Cathode Ray Tube display is used to good effect as a narrowband 'monitor' to show you what's going on up to 100kHz either side of your tuned frequency. Icom's CI-V remote control system is available, with professional software also available.

211

Icom IC-R20

Type;   Handheld
Coverage;   0.150-3305
Modes;   AM, FM, WFM, SSB,
CW

Channels;   1250
A dual watch receiver, which allows you to receive 2 channels simultaneously,
a digital audio recorder for up to 260 minutes, CTCSS and DCS squelch, a
bandscope, and optional PC remote
upload/download.

Icom IC-R5

Type;   Handheld
Coverage;   0.150-1310
Modes;   AM, FM, WFM
Channels;   1250
Including a CTCSS and DCS tone
squelch, and PC upload/download
capability via USB cable.

Icom IC-R3

Type;   Handheld
Coverage;   0.495-2450.095
Modes;   AM, FM,WFM
Video FM,   Video AM
Channels;   400

This unique handheld also has video reception
facilities, although it only handles half of the
2400 MHz band commonly used and from my
tests is rather insensitive for video reception in
this range. 5 and 6.25kHz frequency resolution
but not 8.33kHz for airband.

Icom IC-PCR1500

Type;   Base
Coverage;   0.01-3300
Modes;   AM, FM, WFM, CW, LSB, USB
Channels;   Unlimited
A PC-controlled communications receiver, successor to the IC-PCR1000 and
offers a very wide range of operating facilities, I've used one with good
results and would love to have one permanently in my station!

IC-R1500

Type;   Base / Mobile
Coverage;   0.01-3300
Modes;   AM, FM, WFM,
                 CW, LSB, USB
Channels;   Unlimited

Identical to the IC-PCR1500 and retaining all the same PC-controlled facilities, but with the addition of a manual remote control head for local control without the need of a PC.

Kenwood RZ-1

Type: Base/mobile
Coverage: 500kHz 905MHz
Modes: AM, FM, Wide FM
Channels: 100
Sensitivity: 20MHz 1.7uV AM 145MHz 5.01uV AM 0.28uV FM
                   435MHz 1.09uV FM
Adjacent Channel: 12.5kHz 34.5dB 25kHz 40.5dB
Intermodulation: 50/100kHz 64.0dB

This car-radio-sized scanner covers from Long Wave upwards and with its wideband coverage including HF it could make it a 'different' sort of car radio to have. It's a little insensitive on AM with no squelch on this mode, also to listen to FM stereo you'll need an external stereo amplifier. Very rarely found

nowadays but occasionally available on the secondhand market.

Maycom AR-108

Type;   Handheld
Coverage;   108-136(AM), 136-180(FM)
Modes;   AM FM (auto-selected)
Channels;   198

An economic scanner that's bound to appeal to VHF airband and marine enthusiasts. This model has also been seen under other trade names.

Netset PRO-44

Type:	Handheld
Coverage:	68 88, 108 137, 137 174, 380 512MHz (AM)
Modes:	FM, AM on Airband
Channels:	50
Sensitivity:	145MHz 0.44uV 435MHz 0.70uV
Adjacent Channel:	12.5kHz 11.4dB 25kHz 69.3dB
Blocking 100k	Hz 61.4dB 1MHz 85.7dB 10MHz 86.5dB
Intermodulation:	25/50kHz 56.3dB 50/100kHz 56.9dB

It's cheap, it's remarkably similar to the earlier but more expensive Realistic PRO-43 sold by the same chain of retailers, and like its predecessor it's also available in the high street. It isn't designed specifically for UK use, on VHF you're stuck with 5kHz steps and no AM apart from on the airband section, and on UHF you'll hear signals on the 'image' frequency (i.e., not the one you're tuned to) twice as strong as the one you want to listen to. But it's cheap, and it's easily available.

Netset PRO-46

Type:	Handheld
Coverage:	68 88, 108 137, 137 174, 406 512, 806 956MHz
Modes:	FM, AM on Airband
Channels:	100
Sensitivity:	145MHz 0.26uV 435MHz 0.24uV 934MHz 0.48uV
Adjacent Channel:	12.5kHz 13.3dB 25kHz 64.4dB
Blocking	100kHz 63.2dB 1MHz 78.6dB 10MHz 88.9dB
Intermodulation:	25/50kHz 60.3dB 50/100kHz 59.1dB

This one's rather similar in terms of coverage to its 'smaller brother' the Netset PRO-44, but adds coverage of 806-956MHz, with small 'missing segments' corresponding to the cellular frequency bands used in the US (but not in the UK). This is a 'dead giveaway' of its intended market, and again you're stuck with 5kHz steps on VHF and no AM apart from the Airband range. The poor image rejection here is in the 900Mhz region, where it receives 'image' signals slightly stronger than the wanted signal. However, it's readily available in the high street, it's reasonably priced, and that's what'll sell it.

Netset PRO-2032

Type:	Base/Mobile
Coverage:	68 88, 108 137,
	137 174, 380
	512, 806
	96MHz
Modes:	FM, AM on
	Airband
Channels:	200

Another 'high street' scanner, this one's sure to be a popular choice in the UK due to its wide availability. A fast scanning rate of 25 memories per second, or 50 frequencies per second in 'search' mode, should mean you don't miss much while the set's looking around for signals. 5kHz steps on VHF together with AM only on Airband is a limitation, although the 'value for money' aspect makes up for this.

Nevada MS-1000

Type:	Base/Mobile
Coverage:	500MHz
	600MHz, 800
	1300MHz
Modes:	AM, FM, Wide
	FM
Channels:	1000

Having similar features to some of the Yupiteru base/mobile scanners, this one's an 'OEM' model from the UK firm of Nevada. The wideband coverage should make sure you've plenty to listen to, and the UK source may be an advantage in terms of backup.

Optoelectronics Optocom

Type:	PC Controlled
Coverage:	25-550, 760-1300MHz
Modes:	AM, FM, Wide FM
Sensitivity:	25MHz 0.80uV AM 0.27uV FM 145MHz
	0.59uV AM 0.40uV FM 435MHz 0.36uV
	AM 0.23uV FM 1300MHz 0.78uV FM
Adjacent Channel:	12.5kHz 8.4dB 25kHz 59.3dB
Blocking 100kHz	59.1dB 1MHz 77.8dB 10MHz 79.2dB
Intermodulation:	50/100kHz 57.5dB

The US manufacturers of the Optocom PC-controlled receiver make their control protocol openly available so that third parties can also produce software, and such exists which gives an even more flexible alternative to the supplied software. Used with 'Trunk Tracker' software, which is supplied, the Optocom has the capability of tracking Motorola and EDACS (but not MPT1327) trunked radio system, although software for MPT1327 as widely

used in Europe may also
become available. The set
has a built-in store of 28
memory channels as well as
front-panel volume and
squelch knobs, so after
uploading from a PC it can
be used as a 'stand alone'
receiver when needed.

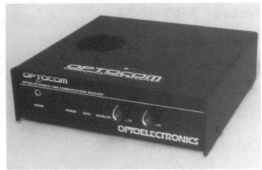

Optoelectronics Scout

Not a scanner as such but a 'frequency finder', it's a sensitive frequency
counter which can hunt out and store active frequencies in use in a local area.
Depending upon the output power and antenna of the transmitter, signals
with ranges of between a few metres and half a kilometre can be detected. The
Scout has a built-in frequency memory to store over 100 active channels,
these can be manually cycled through to view the frequencies. A very useful
facility is a data port which can connect with suitably equipped scanners, e.g.
the AR8000 and 8200, to automatically tune the connected scanner receiver to
the detected frequency.

Optoelectronics Xplorer

This is a wideband sweeping receiver in a self-contained portable metal case,
with a built-in detector and audio amplifier. Like the Scout it detects local
signals, but instead of giving a frequency readout you actually hear the signal
it's found. There's a 'lockout' facility to prevent it halting on unwanted strong
carriers.

Realistic PRO-25

Type:	Handheld
Coverage:	68-88, 108-174, 406-512, 806-956MHz
Modes:	AM Airband, FM
Channels:	100
Sensitivity:	145MHz 0.30uV 435MHz 0.23uV 934MHz 0.26uV
Adjacent Channel:	12.5kHz 9.6dB 25kHz 62.7dB
Blocking 100kHz	65.6dB 1MHz 83.4dB 10MHz 94.9dB
Intermodulation:	25/50kHz 57.5dB 50/ 100kHz 62.1dB

A lightweight portable scanner, although it's
restricted to 5kHz channel steps on VHF and
strangely 12.5kHz steps on VHF airband.

Realistic PRO-26

Type:	Handheld
Coverage:	25-1300MHz continuous
Modes:	AM, FM
Channels:	200
Sensitivity:	145MHz 0.26uV FM,
	435MHz 0.47uV FM,
	934MHz 0.12uV FM
Adjacent Channel:	12.5kHz 0.3dB 25kHz
	51.1dB
Blocking 100kHz	63.5dB 1MHz 69.6dB
	10MHz 88.5dB
Intermodulation:	25/50kHz 55.4dB 50/
	100kHz 56.8dB

This handheld can usefully be switched to either AM or FM throughout it's range, and 10 'monitor' memories in addition to the 200 normal channels are provided for quick storage. A 'default' mode and channel spacing, based on US use, is programmed in which the receiver switches to on each frequency change, although this can be manually changed each time after you've entered the frequency. A triple conversion gives good image rejection, an auto-memory-store facility is fitted which works well, together with a fast channel scan rate of around 50 channels a second.

Realistic PRO-27

Type:	Handheld
Coverage:	68-88, 137-174, 406-512MHz
Modes:	FM
Channels:	20
Sensitivity:	145MHz 0.28uV 435MHz
	0.47uV
Adjacent Channel:	12.5kHz 8.7dB 25kHz 59.5dB
Blocking 100kHz	60.1dB 1MHz 85.9dB
	10MHz 93.0dB
Intermodulation:	25/50kHz 55.7dB 50/100kHz
	61.4dB

A low-cost starter, possibly worth looking at if you're not interested in airband coverage. A couple of button pushes can usefully make the scanner search across one of seven pre-stored band ranges. Easy to use but limited to 5kHz steps on VHF.

Realistic PRO-28

Type: Handheld
Coverage: 68-88, 137-174, 406-
                     512MHz
Modes: FM
Channels: 30
Sensitivity: 145MHz 0.27uV
                     435MHz 0.33uV
Adjacent Channel: 12.5kHz 8.0dB
                     25kHz 56.7dB
Blocking 100kHz   65.3dB 1MHz
                     84.3dB 10MHz
                     93.1dB
Intermodulation: 25/50kHz 52.1dB
                     50/100kHz 60.3dB

Another low-cost non-airband 'starter', with 30 memory channels and 7 pre-stored search ranges although again with 5kHz steps on VHF.

Realistic PRO-29

Type: Handheld
Coverage: 68-88, 108-174, 406-
                     512, 806-965MHz
Modes: AM Airband, FM
Channels: 60
Sensitivity: 145MHz 0.333uV
                     435MHz 0.48uV
                     950MHz 0.39uV
Adjacent Channel: 12.5kHz 5.0dB
                     25kHz 61.2dB
Blocking 100kHz   70.7dB 1MHz
                     86.0dB 10MHz
                     94.4dB
Intermodulation: 25/50kHz 54.0dB
                     50/100kHz 61.7dB

Yet another 'starter' handheld from the high street, this time with VHF Airband included as well as upper UHF. A few extra facilities are included, such as pa programmable search range and 'priority' channel scan, as well as the facility to enter up to 30 'search skip' frequencies for the set to ignore on subsequent searches. The receiver has a fixed step size of 5kHz on VHF.

Realistic PRO-39

Type: Handheld
Coverage: 68-88, 108-137, 137-174, 806-956MHz
Modes: FM, AM on Airband
Channels: 200

A 10 channel 'monitor bank', which you can use as a 'scratch pad' when scanning, helps you fill the set's memory channels adds to the 200 channels. This one's reasonably popular due to its wide availability, although the lack of switchable AM/FM can cause limitations.

Realistic PRO-41

Type: Handheld
Coverage: 68 88, 137 174, 406 512MHz
Modes: FM
Channels: 10

Having just 10 channels which you manually program, this is a low-cost scanner that's available at an economic price. It doesn't have a 'search' facility, so you need to know which frequencies you want to listen to before you can listen to anything at all. But this type of scanner (under the Bearcat BC-50 title plus one or two others) have been quite popular amongst users such as Marine Band listeners who just want to 'keep an ear open' on a few channels.

Realistic PRO-43

Type: Handheld
Coverage: 68 88, 118 174, 220 512, 806
                1000MHz
Modes: AM, FM
Channels: 200
Sensitivity: 145MHz 0.26uV AM 0.13uV FM
                435MHz 0.83uV AM 0.43uV FM
                934MHz 0.55uV AM 0.28uV FM
Adjacent Channel: 12.5kHz 7.0dB 25kHz 33.3dB
Blocking 100kHz 44.5dB 1MHz 71.2dB 10MHz
                94.0dB
Intermodulation: 50/100kHz 49.7dB

Marketed as a 'high performance' scanner, this was the first handheld model from Realistic to have switchable AM and FM across its frequency coverage range. No longer do you have to 'put up' with AM on Airband only from high-street handheld scanners. A fast scanning facility together with 10 'monitor' memories in addition to the 'normal' memory channels make it quite powerful in use. The set looks smart, it's easy to use, and as well as being very sensitive on VHF (to pick up weak signals), it also has reasonable built-in protection against out-of-band signals which many scanners fall down on badly.

Realistic PRO-62

Type:	Handheld
Coverage:	68-88, 118-174, 380-512, 806-960MHz
Modes:	AM, FM
Channels:	200
Sensitivity:	145MHz 0.22uV 435MHz 0.37uV 935MHz 0.46uV
Adjacent Channel:	12.5kHz 6.3dB 25kHz 56.9dB
Blocking 100kHz	67.3dB 1MHz 77.7dB 10MHz 93.0dB
Intermodulation:	25/50kHz 52.5dB 50/100kHz 52.3dB

A reasonably lightweight handheld powered from 6 AA cells, with AM or FM programmable on any frequency as well as 5kHz, 12.5kHz or 25kHz channel steps for search mode, although 'default' mode and channel steps based on US use is initially selected when you change frequency.

Realistic PRO-63

Type:	Handheld
Coverage:	68-88, 118-174, 380-512MHz
Modes:	AM Airband, FM
Channels:	100
Sensitivity:	145MHz 0.22uV 435MHz 0.27uV
Adjacent Channel:	12.5kHz 8.4dB 25kHz 64.7dB
Blocking 100kHz	74.9dB 1MHz 85.1dB 10MHz 87.6dB
Intermodulation:	25/50kHz 58.2dB 50/100kHz 57.8dB

This one's described as a portable 'event scanner' due to its coverage range, as it could indeed be useful when you're out at the air show or yacht race. Fixed search banks acts as an easy-to-use lookout for active channels, although again fixed 5kHz steps are used on VHF.

Realistic PRO-70

Type:	Handheld
Coverage:	68-88, 137-174, 380-512MHz
Modes:	AM, FM
Channels:	50
Sensitivity:	145MHz 0.32uV 435MHz 0.23uV
Adjacent Channel:	12.5kHz 8.7dB 25kHz 56.8dB
Blocking 100kHz	58.1dB 1MHz 95.0dB 10MHz 93.5dB
Intermodulation:	25/50kHz 52.5dB 50/100kHz 52.3dB

Easy to use, FM-only scanner with 5kHz steps on VHF and 12.5kHz steps on UHF and nine pre-stored search banks, powered from six AA cells.

Realistic PRO-2006

Type:	Base/mobile
Coverage:	25 520, 760 1300MHz
Modes:	AM, FM, Wide FM
Channels:	400
Sensitivity:	29MHz 0.22uV FM 145MHz 1.08uV FM 435MHz 0.39uV FM 934MHz 0.48uV FM
Adjacent Channel:	12.5kHz 27.5dB 25kHz 50.3dB
Blocking 100kHz	69dB 1MHz 88dB 10MHz 94dB
Intermodulation:	50/100kHz 66.5dB

From the number of these being 'snapped up' by scanner purchasers when it first came out, this set looks like it's one of the most popular base scanners on the UK market. Switchable AM and FM across its coverage range gives it that bit more 'usefulness' in the UK, and unlike earlier Realistic models you can choose step sizes of 5kHz, 12.5kHz, or 50kHz. Although it's not one of the cheapest sets, it works well on the air, and the wide frequency coverage is likely to make the set rival with other 'up market' scanners available from specialist dealers.

Realistic PRO-2014

Type: Base
Coverage: 68-88, 137-174, 380-512MHz
Modes: FM
Channels: 50
Sensitivity: 145MHz 0.65uV FM 435MHz
                     0.37uV FM
Adjacent Channel: 12.5kHz 7.4dB 25kHz 62.8dB
Blocking 100kHz   66.3dB 1MHz 91.7dB 10MHz
                     94.2dB
Intermodulation: 25/50kHz 56.5dB 50/100kHz
                     56.8dB

Basically a base station model of one
of Realistic's easy-to-use FM
handhelds, not unlike the PRO-70 in
facilities, with ten pre-stored search
bank and 65kHz steps on VHF, 12.5kHz
on UHF. A telescopic whip is supplied,
the external antenna connector being a
Motorola 'car radio' type.

Realistic PRO-2035

Type: Base
Coverage: 25-520, 760-1300MHz
Modes: AM, FM, Wide FM
Channels: 1000
Sensitivity: 25MHz 0.21uV FM 145MHz 0.41uV FM 435MHz 0.26uV
                     FM 934MHz 0.25uV FM
Adjacent Channel: 12.5kHz 0.7dB 25kHz 24.2dB
Blocking 100kHz   55.4dB 1MHz 78.2dB 10MHz 88.6dB
Intermodulation: 25/50kHz 58.7dB, 50/100kHz 56.7dB

Together with keypad control, a rotary tuning knob also acts as a channel and
frequency change for home use. A very fast scan rate of 50 channels per second,
although the auto-store repeats previously found channels. Channel steps are
selectable between 5, 12.5 and 25kHz on any range, as well as selectable AM, FM
and WFM modes.

Realistic PRO-2036

Type: Base
Coverage: 66-88, 108-174, 216-512, 806-956MHz
Modes: AM Airband, FM
Channels: 200
Sensitivity: 145MHz 0.34uV 435MHz 0.26uV 950MHz 0.23uV FM
Adjacent Channel: 12.5kHz 7.6dB 25kHz 60.8dB
Blocking 100kHz   75.5dB 1MHz 84.0dB 10MHz 86.7dB
Intermodulation: 25/50kHz 55.9dB 50/100kHz 63.4dB

A neatly-styled base station scanner with the usual handheld facilities but with built-in CTCSS decode capability, which can also be used in scan mode to only halt and enable the speaker when the correct CTCSS tone is present on that channel. Realistic have overcome the PRO-2035's repeat store in 'auto-store' memory mode with this one, as it works very well in the PRO-2036.

Realistic PRO-2045

Type: Base
Coverage: 68-88, 108-174, 216-512, 806-1000MHz
Modes: AM, FM
Channels: 200
Sensitivity: 145MHz 0.37uV 435MHz 0.26uV 950MHz 0.55uV
Adjacent Channel: 12.5kHz 7.7dB 25kHz 51.0dB
Blocking 100kHz   51.9dB 1MHz 82.6dB 10MHz 96.6dB
Intermodulation: 25/50kHz 61.5dB 50/100kHz 62.3dB

This one has useful base station monitoring facilities such as a 'hit count' mode to see how many times a channel has been active in your absence, a switchable attenuator, switchable AM and FM on any frequency, an auto-store that works well, and a search speed of an incredible 300 steps per second. Up to 50 channels can be programmed to be skipped on subsequent searches, but you're still limited to 5kHz steps on VHF though, unusually with 12.5kHz rather than 25kHz steps on VHF airband.

Realistic PRO-9200

Type: Base
Coverage: 68-88, 108-174, 406-512MHz
Modes: AM Airband, FM
Channels: 16
Sensitivity: 145MHz 0.43uV 435MHz 0.41uV
Adjacent Channel: 12.5kHz 8.5dB 25kHz 63.5dB
Blocking 100kHz   76.5dB 1MHz 90.0dB 10MHz 96.5dB
Intermodulation: 25/50kHz 67.0dB 50/100kHz 65.5dB

An uncomplicated and easy to use scanner for the home, switching on immediately starts it off scanning thorough its 16 memory channels. There's also a search mode for finding new frequencies, with 5kHz steps on VHF and 12.5kHz steps on UHF, 25kHz on airband. A car radio type antenna connector is used for the supplied telescopic whip.

Shinwa SR-001

Type: Base/mobile
Coverage: 25 1000MHz
Modes: AM, FM, Wide FM
Channels: 200
Sensitivity: 25MHz 5.20uV AM 2.51uV FM 145MHz 0.68uV AM
                      0.36uV FM 435MHz 1.07uV AM 0.43uV FM 934MHz
                      4.32uV AM 2.15uV FM
Adjacent Channel: 12.5kHz 26.0dB 25kHz 53.0dB
Blocking 100kHz   76.5dB 1MHz 88.5dB 10MHz 96.5dB
Intermodulation: 25/50kHz 50.5dB 50/100kHz 68.8dB

This one looks like it was designed to be a hot contender for the 'alternative car radio' market, although the manufacturer's choice of including a TV/Video style remote control with the radio is rather puzzling. Despite its features, this set doesn't quite seem to have 'caught on'. Not a commonly found scanner but you might occasionally come across one second hand.

Signal R535

Type: Base/mobile/transportable
Coverage: 108 143, 220 380MHz
Modes: AM
Channels: 60

A specialised airband receiver with options available to enable it to be used while being carried around. A fairly tedious type of programming is involved but the set is highly spoken of by airband enthusiasts. Not often found nowadays but if you find one second hand don't dismiss it.

Skyscan 747

Type;   Handheld
Coverage;   522-1629 kHz 87.3-108.1 MHz,
                    Marine Channels 1-28,
                    37,39,60-74 &77-88
Modes;   AM, WFM, FM (auto-
                    selected)
Channels;   10

More of a domestic AM/FM receiver but with the added bonus of VHF marine channels, and very keenly priced.

Sony Air-7

Type: Handheld
Coverage: 100kHz 2.2MHz, 76 136MHz
Modes: AM, FM, Wide FM
Channels: 30

Bulky, heavy, limited coverage and expensive. However, airband fans say it performs superbly and the AGC (very important with AM) is excellent.

Sony PRO-80

Type: Handheld
Coverage: 150MHz 108MHz, 115 223MHz
Modes: AM, FM, SSB
Channels: 40
Similar in shape, size, and weight to the Air-7 and also expensive. Although the PRO-80 performs well, is built to Sony's usual high standard, and includes LW, MW, and full HF coverage, it does not include UHF. The VHF coverage is provided by a plug-in adapter, which seems to be an afterthought on the part of the designer.

Standard AX-400

Type:	Portable
Coverage:	500kHz 1300MHz
Modes:	AM, FM, Wide FM
Channels:	400
Sensitivity:	20MHz 0.34uV AM 145MHz 0.15uV FM 435MHz 0.24uV FM 934MHz 0.15uV FM
Adjacent Channel:	12.5kHz 24.8dB 25kHz 54.1dB
Blocking: 100kHz	54.0dB 1MHz 83.2dB 10MHz 94.0dB
Intermodulation:	25/50kHz 52.6dB 50/100kHz 55.8dB

A tiny wideband scanner, very easily portable, and powered by two AA sized batteries. Standard seem to have a knack of squeezing circuitry into a small case without sacrificing on performance. The use of a standard BNC connector allows external mobile and portable antennas to be easily connected. I use one of these scanners myself as a top-pocket portable when I'm out and about. The Welz WS-1000 scanner is virtually identical.

Standard AX-700

Type:	Base/mobile
Coverage:	50 905MHz
Modes:	AM, FM, Wide FM
Channels:	100
Sensitivity:	145MHz 0.39uV AM 0.18uV FM 435MHz 0.26uV AM 0.25uV FM 905MHz 0.54uV AM 0.36uV FM
Adjacent Channel:	12.5kHz 31.0dB 25kHz 65.3dB
Blocking 100kHz	73.0dB 1MHz 78.0dB 10MHz 100dB
Intermodulation:	50/100kHz 58.5dB

If you fancy keeping an eye on what's going on above and below the channel you're tuned to, as well as listening to what you've tapped in on the keypad, this one's for you. It has a panoramic display in the form of a LCD bargraph, showing signal levels across a frequency range of 1MHz, 250kHz, or 100kHz. Some user might find the 905MHz upper frequency limiting.

Uniden UBC-3500XLT

Type;   Handheld
Frequency range;   25-512MHz, 806 - 1300MHz
Mode;   FM, FMB, WFM, AM
Channel steps;   5, 6.25, 8.33, 7.5, 10, 12.5, 15,
                    20, 25, 50, 100 kHz
Memory channels;   2500 (dynamic)
Powered by 3 x AA batteries
Another scanner with 'Close Call' facility for automatically searching out and storing strong local signals in the vicinity. Sub-tone and DCS squelch including Sub-tone and DCS scan. Usefully includes 8.33kHz steps for VHF airband, although no military airband coverage

Uniden-Bearcat UBC-280XLT

Type;   Handheld
Frequency Range;   25-88MHz, 108-512MHz,
                    806-956MHz
Modes;   NFM, AM
Channel steps   5, 12.5kHz
Aimed primarily at US sporting enthusiasts, this handheld comes with 10 pre-programmed search bands for such US frequencies. It usefully has sub-tone and DCS search facilities, but channel steps of 5kHz and 12.5kHz limit its usefulness in the UK.

Uniden-Bearcat ScanCat 230

Type;   Handheld
Frequency range;   25-54, 108-174, 216-225, 400-
                    512, 806-956, 1240-1300MHz
Memories;   2500 (dynamic)
Powered by; 2 x AA batteries
A 'Close Call' scanner (for automatic monitoring of local signals) that's again aimed at US sporting enthusiasts although this one is rather better value than other close call types.

Uniden-Bearcat UBC-69XLT-2

Type;   Handheld
Frequency range:;   25 - 88MHz, 137 - 174MHz, 406 - 512MHz
Modes;   FM
Memories;   80
Channel steps;   5, 6.25, 10, 12.5kHz
Powered by 2 x AA batteries
A stylish, easy to use and economic scanner, worth considering if you're not interested in airband coverage. Useful channel steps of 5, 6.25, 10 and 12.5kHz are provided

Watson Super Searcher

A self-contained and portable frequency hunter which displays the frequency of signals it finds, in a similar manner to the Optoelectroniucs 'Scout', but without all the memories. With its data port it can also usefully connect with a suitably equipped scanner such as the AOR AR-8000 to automatically tune the set to the detected frequency.

Welz WS-1000

This is virtually identical to the Standard AX-400, detailed earlier.

WIN 108

Type: Handheld
Coverage: 108 143MHz
Modes: AM
Channels: 20
A scanner that has received mixed reactions. Some owners speak highly of it but some reviews have been critical with claims that the keyboard is flimsy and difficult to operate and sensitivity could be better on a set designed solely for AM mode.

Winradio

Type: PC Controlled
Coverage: 50kHz-1300MHz
Modes: AM, FM, Wide FM, SSB
Sensitivity: 20MHz 3.85uV SSB 0.46uV AM 145MHz 0.86uV
                      SSB 0.45uV AM 0.46uV FM 435MHz 0.29uV SSB
                      0.37uV AM 0.36uV FM 1300MHz 3.51uV SSB
                      2.39uV FM
Adjacent Channel: 12.5kHz 15.8dB 25kHz 45.5dB
Blocking 100kHz   48.8dB 1MHz 76.9dB 10MHz 86.0dB
Intermodulation: 25/50kHz 59.2dB 50/100kHz 57.2dB
This comes as a plug-in PC card, fitting internally to your PC. On-

screen tuning and memory facilities are controlled via the PC mouse and keyboard with unlimited memory channels and an auto-store facility. There's also a useful memory 'search' where you can find memory channels by entering a partial text string. HF performance is naturally limited to that of a typical scanner rather than a purpose-designed HF receiver.

Yaesu FRG-9600

Type: Base/mobile
Coverage: 60-950MHz
Modes: AM, FM, Wide FM, SSB
Channels: 100
One of the earliest non-Bearcat base station scanners from Japan, and many are around second hand, being a very popular scanner in the early days. Some models have the advantage of a UK-fitted HF converter, useful with the included SSB mode, as well as extended UHF coverage above 905MHz. The scan mode always continues 10 seconds after finding a signal which is limiting to some users, although there is a CAT computer control port on the rear panel which extends the receiver's versatility.

Yaesu VR120-D

Type;   Handheld
Coverage;   0.1-1300
Modes;   AM, FM, WFM
Channels;   640
A rugged scanner capable of taking the 'knocks' in life, with an easy-to-use one-touch memory system for high priority stations. The 'D' version (VR120-D) adds a charge socket.

Yaesu VR500

Type;	Handheld
Coverage;	0.1-1300
Modes;	AM, FM, WFM, CW, LSB, USB
Channels;	1000

A compact handheld with all-mode
reception and wide frequency coverage.

Yaesu VR5000

Type;	Base
Coverage;	0.1-1300
Modes;	AM, Narrow AM, Wide AM, FM, WFM, CW, LSB, USB
Channels;	1000

A high-performance and versatile desktop receiver, offing many of the
facilities of top-flight transceivers. I've used one myself to very good effect.

Yupiteru VT-125

Type:	Handheld
Coverage:	108-142MHz
Modes:	AM
Channels:	30
Sensitivity:	118MHz 0.35uV 125MHz 0.34uV 136MHz 0.34uV
Adjacent Channel:	25kHz 60.8dB 50kHz 66.3dB
Blocking 100kHz	70.0dB 1MHz 89.0dB 10MHz 84.5dB
Intermodulation:	25/50kHz 50.5dB 50/100kHz 50.0dB

This is a 'cut down' version, dedicated for Civil
Airband listening, of Yupiteru's other 'do everything'
handheld scanners. It's inexpensive and quite light
and easy to carry around. Watch out for the narrow
filtering, as this could distort some of the
(deliberately) offset signals from land based Airband
transmitters. The available 30 memory channels could
be limiting for active airband enthusiasts.

Yupiteru VT-150

Type:	Handheld
Coverage:	142MHz 170MHz
Modes:	FM
Channels:	100

The dedicated 'Marine Band' cousin to the Airband VT-125 handheld scanner. Again it looks very much like a 'cut down' version of Yupiteru's 'do-everything' handheld scanners, but again without the higher price tag. A possible choice for the user interested primarily in Marine Band monitoring, although the 30 channels could be a bit limiting.

Yupiteru VT-225

Type:	Handheld
Coverage:	108-142.1, 149.5-160, 222 391MHz
Modes:	AM, FM
Channels:	100
Sensitivity:	130MHz 0.29uV AM 155MHz 0.16uV FM 250MHz 0.32uV
Adjacent Channel:	12.5kHz 48.3dB 25kHz 52.0dB
Blocking 100kHz	63.5dB 1MHz 91.0dB 10MHz 94.5dB
Intermodulation:	25/50kHz 53.0dB 50/ 100kHz 62.5dB

This set from the Yupiteru collection looks like it's meant for Airband (Civil and Military) and Marine Band enthusiasts, a bit of a 'mixture' in fact. They obviously feel there's a

'niche' market for such a set, and I believe they could well find it amongst users wanting a lightweight scanner to carry to air shows and the like.

Yupiteru MVT-3100

Type:	Handheld
Coverage:	143-162.025, 347.7125-452, 830-960MHz
Modes:	FM
Channels:	100
Sensitivity:	145MHz 0.21uV FM 435MHz 0.18uV FM 934MHz 0.35uV FM
Adjacent Channel:	12.5kHz 37.6dB 25kHz 45.7dB
Blocking 100kHz	62.8dB 1MHz 77.4dB 10MHz 69.3dB
Intermodulation:	25/50kHz 61.7dB 50/100kHz 59.5dB

This one's another 'niche market' set from Yupiteru, although quite what 'niche', I'm not too sure. It's 'hard programmed' with 10kHz steps over 143 155MHz and 430 440MHz, this includes the 2m and 70cm amateur bands, which instantly cuts out half of all the 25kHz channels used. However, I found it to be a very good performer on 156MHz Marine Band when I used it on the water, and the set very rapidly steps through programmed memory channels 30 at a time. A further useful feature is that it has up to 100 'pass' frequencies, which you can automatically program to skip with the set in 'search' mode.

Yupiteru MVT3300

Type;   Handheld
Coverage;   66-88, 108-174, 300-470, 800-1000
Mhz Modes;   AM/FM
Channels;   200

A budget scanner yet with quite reasonable performance.

Yupiteru MVT-7000

Type: Handheld
Coverage: 1 1300MHz continuous
Modes: AM, FM, Wide FM
Channels: 200
Sensitivity: 29MHz 0.18uV FM 145MHz 0.24uV FM 435MHz 0.24uV
                  FM 934MHz 0.25uV FM
Adjacent Channel: 12.5kHz 17.0dB 25kHz 47.0dB
Blocking 100kHz   61.0dB 1MHz 84.0dB 10MHz 89.0dB
Intermodulation: 50/100kHz 52.0dB

The tuning range of this set covers down to 100kHz with reduced sensitivity, so as well as being a 'listen to everything on VHF/UHF' you can also tune into Medium Wave and HF broadcast stations for that bit of 'alternative' listening. This one's a competitor in terms of frequency coverage to the IC-R1 and DJ-X1, and although it doesn't have the small size of its competition it doesn't have the poor performance of the others in terms of strong signal handling either. You pays your money and takes your choice.

Yupiteru MVT-7100

Type: Handheld
Coverage: 1 1300MHz continuous
Modes: AM, FM, Wide FM, LSB, USB
Channels: 1000
Sensitivity: 20MHz 0.16uV SSB 0.22uV AM
                    0.13uV FM 145MHz 0.18uV SSB
                    0.25uV AM 0.16uV FM 435MHz
                    0.26uV SSB 0.35uV AM 0.16uV
                    FM 934MHz 0.33uV SSB
                    0.37uV AM 0.21uV FM
Adjacent Channel: 12.5kHz 34.1dB 25kHz 48.4dB
Blocking: 100kHz   55.5dB 1MHz 85.5dB 10MHz
                    93.8dB
Intermodulation: 25/50kHz 65.3dB 50/100kHz
                    24.0dB

The very wide frequency coverage, together with 'real' SSB reception and the ability to tune (and store frequencies) on SSB in 50Hz steps, has made this a very sought-after set amongst scanner devotees. I've used one on many occasions, and I've always been sad to give it back, it just looks like I'll have to buy one of these when I've saved up. About the only thing I don't like about it is the telescopic whip, which will surely break in use, but at least this lets you adjust its length to 'peak' on the part of the set's very wide frequency range you're listening to at any given time.

Yupiteru MVT-7200

Type: Handheld
Coverage: 500kHz-1650MHz
                    continuous
Modes: AM, FM, Wide FM, LSB,
                    USB
Channels: 1000
Sensitivity: 20MHz 0.16uV
                    SSB 0.22uV AM 0.13uV
                    FM 145MHz 0.18uV SSB
                    0.25uV AM 0.16uV FM
                    435MHz 0.26uV SSB
                    0.35uV AM 0.16uV FM
                    934MHz 0.33uV SSB
                    0.37uV AM 0.21uV FM
Adjacent Channel: 12.5kHz 34.7dB 25kHz
                    46.5dB
Blocking: 100kHz   66.5dB 1MHz 86.2dB
                    10MHz 94.6dB
Intermodulation: 50/100kHz 65.1dB

This is the physically similar successor to the MVT-7100 with almost similar facilities but an extended frequency coverage and significantly better IF filtering for SSB reception, also an added 'narrow AM' bandwidth which is especially useful on the crowded HF bands. The easily breakable telescopic has also been sensibly replaced with a helical whip.

Yupiteru MVT7300

Type;   Handheld
Coverage;   531kHz-1320MHz
Mhz Modes;   AM, FM, WFM, NAM, USB, LSB, CW
Channels;   1000

A multimode scanner with plenty of operating facilities

Yupiteru MVT-8000

Type: Base/mobile
Coverage: 8 1300MHz continuous
Modes: AM, FM, Wide FM
Channels: 200

Essentially a 'base/mobile' version of the Yupiteru MVT-7000, adding a switchable attenuator to help with strong signals from external antennas. It offers a wide frequency coverage in a very small case, which should fit quite neatly in your car as well as presenting a smart 'low profile' on a desktop.

Yupiteru MVT-9000

Type: Handheld
Coverage: 512kHz-2039MHz
                   continuous
Modes: AM, FM, Wide FM, LSB,
                   USB, CW
Channels: 1000
Sensitivity: 20MHz 0.09uV SSB
                   0.24uV AM 0.15uV FM
                   145MHz 0.15uV SSB
                   0.36uV AM 0.23uV FM
                   435MHz 0.22uV SSB
                   0.29uV FM 950MHz
                   0.30uV SSB 0.44uV FM
Adjacent Channel: 12.5kHz 39.7dB 25kHz
                   51.5dB
Blocking: 100kHz   68.7dB 1MHz 80.4dB
                   10MHz 83.5dB
Intermodulation: 50/100kHz 65.1dB

Yupiteru's top of the range handheld, with both an external plug-in whip via a BNC connector and a selectable internal ferrite rod antenna for LF band portable reception. A dual frequency display, band-scope, and plenty of scanning and searching modes make this a very desirable and powerful handheld scanner. A built-in switchable speech inverter demodulates simple forms of scrambled transmissions.

Yupiteru MVT9000MKII

Type;   Handheld
Coverage;   530kHz-2039MHz
Modes;   AM, FM, WFM, USB, LSB, CW
Channels;   1000

One of my personal favourites (I own one), includes a 'real time' bandscope, a descrambler for inverted audio transmissions, twin receiver, the business. Only thing missing is 8.33kHz steps for airband.

Antennas

Besides the usual antennas of dipoles, whips, discone and so on, detailed here are a few specialist antennas, which may be of interest to scanner users.

Product: Radac

Manufacturer: Revco (available from Garex).

Description: This is what is known as a 'Nest of dipoles' type of antenna. It is quite an old idea which has been re-vamped to meet the needs of scanner users. In theory it provides reception on six bands which are determined by the length of each of the individual dipole sections. The manufacturers say that for those six bands, antenna gain will be better than a discone. Elements can be anywhere in the range 25-500MHz.

Product: Create CLP5130-2

Type: Log periodic.

Description: A 20 element Log periodic antenna with coverage from 105MHz to 1300MHz. The gain is quoted as 11-13dB with a front to back ratio of 15dB. The antenna has a width of 1.4 metres and is 1.4 metres long. VSWR across the range is quoted as 2:1 and the termination is via an N-socket.

Product: Create CLP5130-1.

Type: Log periodic.
Description: A wider band version of the above antenna covering 50-1300MHz. A 24 element unit which is 3 metres wide and 2 metres long, otherwise the specifications are similar.

Product: Diamond D707.

Type: Multiband pole.
Description: The antenna consists of a slim pole 95cms long which contains a broadband 20dB gain signal pre-amplifier. However, no details are given of the type of elements employed nor the performance across the range which is quoted as 2-1500MHz. The antenna is supplied with a small power unit which sends voltage for the pre-amplifier up the coaxial feeder cable and some provision is made in the interface to vary the gain of the system. Available from Waters and Stanton.

Product: Diamond D505.

Type: Multiband mobile antenna.
Description: Essentially a mobile version of the above antenna with the same specifications. The unit consists of a mobile mounting whip with two loading coils and built-in pre-amplifier. The antenna is 80cms long. Available from Waters and Stanton.

Replacement whips

About the most-commonly damaged part of a portable scanner is, you've guessed it, the set-top whip. Dealers such as Nevada and Waters & Stanton can supply a range of plug-in antennas for portable hand-helds, including UHF and airband helicals.

Tunable Antenna Filter

Many of the scanners detailed in this book have an enormous frequency coverage in a very small package. But unfortunately selectivity and strong signal handling characteristics have sometimes been sacrificed to save space and cost, which means that unwanted signal breakthrough can be a big problem. Attaching a microscopic handheld to a rooftop antenna is often a good recipe for disappointment. One remedy from strong local signals but in a different band is to use a notch filter, which is a high-Q tuned circuit plugged in line with the antenna and can be adjusted to attenuate an unwanted signal. One type, covering the 85-175MHz range is marketed by Garex Electronics. It simply fits in line with the antenna feeder, and lets you tune unwanted signals out. Provided the interference is spaced more than 10MHz away, there's little difference in the signal you want to hear.

Mobile Scanner Antennas

If you're using your handheld scanner on the move, then a suitable

antenna fitted to the outside of the car can make a tremendous difference. 'Putting up' with the set-top antenna is OK as a temporary measure, but you won't get the best from your scanner. Purpose-designed mobile scanners will, of course, always need some form of external antenna. Very often, a simple quarter wave whip, cut for the centre frequency of the band you're mainly interested in (such as Civil Airband), can make a reasonable 'all-round' antenna for mobile scanner use. See Chapter 5 for details of lengths needed. Wideband antennas are a different matter though. A number of multi-

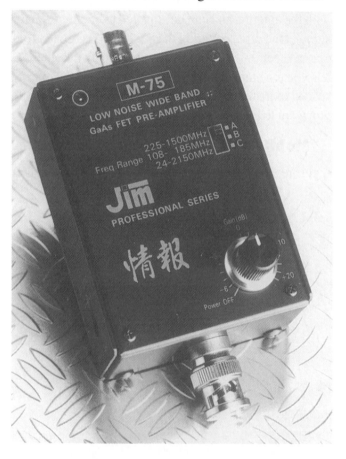

band antennas are available from amateur radio and scanner dealers, including wideband amplified types. A handy tip is that a 'dual band' whip for the 2m and 70cm amateur bands makes a good all-round VHF/UHF mobile scanner antenna. If you'd prefer not to drill holes or clip antenna mounting brackets to your car gutter or boot lip, then a glass-mounted wideband antenna could be useful. A 'glassmount'

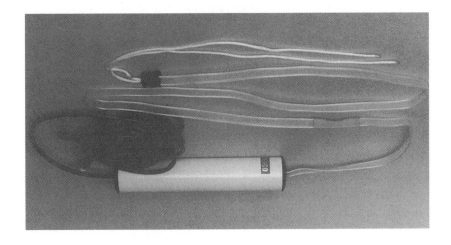

antenna sticks onto one of the windows of your car, usually the rear windscreen, using the glass as a 'dielectric' between the inner and outer fittings. Waters and Stanton Electronics distribute what could be an ideal antenna, which is designed for wideband scanner receive-only use over 30-1200MHz. This is the Pro-Am TGSBNC, and it comes with everything you need apart right down to the BNC plug at the end of the length of coax. The antenna element itself can be unscrewed from its base for carwashes or security against other damage when not in use. If you're fitting one of these, make sure you get the position 'right first time' (check the travel of windscreen wiper blade) as it's very difficult, if not impossible, to change once it's stuck! A re-mounting kit is, however, available should you change cars.

Antenna Amplifiers

There are a number of wideband antenna preamps available, from firms such as Solid State Electronics and Garex. Garex for example produce the GA-4M GAsFET amplifier covering 20-1000MHz, and also produce a low cost VHF Airband preamplifier which is designed to cover 118-137MHz with strong out-of-band signals attenuated. SSE supply the 'Jim' series of preamplifiers which can be switched to either wideband or narrow, together plus other accessories such as handheld base and mobile mounts and chargers.

Indoor Scanner Antennas

If circumstances confine you to using indoor antenna, then try at least to get this near to a window, you'll usually notice a big improvement. A popular portable VHF/UHF scanner antenna is the 'Nomad' from Garex, which is available in normal and active amplified versions. It's a lightweight arrangement using ribbon cable elements and comes fitted with 4m of coax and a BNC plug, it's optimised for the VHF airband but works very well across the VHF/UHF range. It's ideal for using your scanner from temporary locations as an improvement to a set-top whip. For VHF and HF reception, e.g. HF/VHF airband, then Garex also manufacture an excellent 'rollup and take it with you' wire antenna which gives good performance on both ranges, it comes complete with all the required hardware etc.

Chapter 11
UK Scanner and Accessory Manufacturers, Distributors and Dealers

Listed below are a selection of UK and Eire based suppliers of scanners and accessories. Inclusion in this list does not suggest any recommendation or otherwise, also any omission of dealers doesn't mean they're not recommended or worthy of inclusion, it's just that the author wasn't aware of their activity at the time of writing. For contact details of other suppliers local to you, just take a look in the yellow pages phone book under "Radio Communications" and "Aerials" or current adverts in specialist hobby radio publications, details of which are also given here.

Aerial Techniques

Suppliers of masts, brackets, rotators and other accessories.
59 Watcombe Road, Southbourne, Bournemouth, Dorset BH6 3LX Tel: 01202 423555
www.aerial-techniques.com/

Air Supply Aviation Store Ltd

Scanners, Aero charts, antennas
97 High St, Yeadon, Leeds LS19 7TA Tel: 0113 250 9581
www.airsupply.co.uk

AOR (UK)

Distributors of AOR scanners, Ten Tec equipment, antennas, and PC

control software for AOR scanners.
Unit 9, Dimple Road Business Centre, Matlock, Derbyshire DE4 3JX, England Tel: 01629 581222
www.aoruk.com

AVA Electrical Ltd

Independent electrical dealer specialising in radio equipment.
22 Station Road, Sheringham, Norfolk, NR26 8RE, Tel. 01263 823220
www.avaelectrical.com

bhi Ltd

Noise-reducing add-on accessories for receivers
PO Box 318, West Sussex, Burgess Hill, RH15 9NR Tel. 0845 217 9926
www.bhinstrumentation.co.uk

Coastal Communications

19 Cambridge Road, Clacton-on-Sea, Essex CO15 3QJ, Tel. 01225 474292
www.coastalcomms.org.uk

Garex Electronics

Scanner accessories, preamps, antennas and filters
PO Box 52, Exeter, EX4 5FD Tel 07714 198374
www.garex.co.uk

Jaycee Electronics Ltd

Scanners and accessories, Waters and Stanton store (see below)
20 Woodside Way, Glenrothes, Fife KY7 5DF, Tel. 01592 756962
www.jayceecomms.com

Javiation

Scanners, antennas, aviation charts, frequency lists, pre-amps,PC interface leads, software.
PO Box 708, Bradford BD2 3XA Tel. 01274 639503
www.javiation.co.uk/

LAM Communications

Scanners, receivers, accessories, part exchange
52 Sheffield Road
Hoyland Common, Barnsley, South Yorkshire S74 0DQ Tel. 01226 361700
www.lamcommunications.net

ML&S Martin Lynch and Sons

Wide range of hobby radio products and accessories
Outline House, 73 Guildford Street, Chertsey, Surrey KT16 9AS, Tel. 01932 567 333
www.hamradio.co.uk

Modern Radio

Specialists in receivers, accessories and components
101-103 Derby Street, Bolton BL3 6HH, Tel. 01204 526916
www.modernradio.co.uk

Moonraker (UK) Limited

Manufacturers of a wide range of antennas, coax, and antenna accessories
Cranfield Road, Woburn Sands, Bucks MK17 8UR Tel. 01908 281705
www.moonrakerukltd.com/

Nevada

Suppliers of a wide range of scanner equipment, antennas, coax, and accessories
Unit 1 Fizherbert Spur, Farlington, Portsmouth PO6 1TT, Tel. 023 9231 3090
www.nevada.co.uk

P&D CB and Amateur Radio

Suppliers of hobby radio equipment.
1 Knockbracken Drive, Coleraine, Co. Londonderry, N. Ireland, BT52 1WN Tel. 028 7035 1335
www.hamradi-ni.com

Poole Logic

Wide range of scanners, receivers, books and accessories
49 Kingston Road, Poole, Dorset BH15 2LR Tel. 01202 683093

QSL Communications

Suppliers of hobby radio equipment and accessories.
Unit 6, Worle Industrial Centre, Coker Road, Worle, Weston-Super-Mare BS22 6BX, Tel. 01934 512757
www.qsl-comms.co.uk

Radioworld

Suppliers of new and used hobby radio equipment
42 Brook Lane, Great Wyrley, Walsall, WS6 6BQ, Tel. 01922 414796
www.radioworld.co.uk

Shortwave Shop Ltd

Specialists in recevers
18 Fairmile Road, Christchurch, Dorset, BH23 2LJ, Tel. 01202 490099
www.shortwave.co.uk

Stewart Aviation

Air band radio specialists, suppliers of equipment and books
PO Box 7, Market Marborough, Leics LE16 8YL Tel. 01536 770962
www.stewart-aviation.co.uk

Tennamast Scotland Ltd
Antenna support equipment
81 Mains Road, Beith, Ayrshire KA15 2HT Tel 01505 503824
www.tennamast.com

Waters and Stanton PLC
Suppliers of a wide range of scanner equipment, antennas, coax, and accessories
Spa House, 22 Main Road, Hockley, Essex SS5 4QS (further store also in Glenthroes, Scotland, see Jaycee above). Tel. 01702 204965
www.wsplc.com

National Organisations

AMSAT UK; Amateur Satellite Organisation, a voluntary non-profit-making organisation dedicated to the furtherance of amateur radio satellites. They publish an excellent bi-monthly newsletter. Membership information available by sending an SAE to; AMSAT-UK, Badgers, Letton Close, Blandford, Dorset DT11 7SS
www.amsat.uk.org
ISWL; International Short Wave League. 20a Poplar Rd, Healing, Grimsby, DN41 7RD, www.iswl.org.uk
Formed in 1946, a group dedicated to the interests of radio listening world-wide, they publish a quarterly newsletter and run a number or scanner and short-wave listening contests each year.
www.iswl.org.uk
RSGB; Radio Society of Great Britain
UK's national group representing radio amateurs and their interests, also publishers radio-related books). 3 Abbey Court, Fraser Road, Priory Business Park, Bedford MK44 3WH Tel. 01234 832700
www.rsgb.org.uk
RIG: Remote Imaging Group; (International Group for Weather Satellite and Fax Reception) who publish a quarterly newsletter and cater for the interests of weather satellite enthusiasts). PO Box 2001, Dartmouth, Devon TQ6 9QN
www.rig.org.uk

Monthly Magazines

PW Publishing Ltd
Publishers of 'Radio User' magazine with monthly scanning articles, as well as 'Practical Wireless' for amateur radio hobbyists, also publishers and retailers of a wide range of scanner and hobby radio books
Arrowsmith Court, Station Approach, Broadstone, Dorset BH18 8PW
Tel. 01202 659920 www.pwpublishing.ltd.uk

Index